S0-BWT-527

A PHILOSOPHICAL ESSAY

This is a volume in the Arno Press collection

History of Geology

See last pages of this volume
for a complete list of titles

A Philoſophical ESSAY

DECLARING
The probable Cauſes, whence
Stones are produced in the
Greater World

Thomas Sherley

ARNO PRESS

A New York Times Company

New York / 1978

Editorial Supervision: ANDREA HICKS

Reprint Edition 1978 by Arno Press Inc.

Reprinted from a copy in
The Yale University Library

HISTORY OF GEOLOGY
ISBN for complete set: 0-405-10429-4
See last pages of this volume for titles.

Manufactured in the United States of America

Publisher's Note: This book has been reproduced from the best available copy.

Library of Congress Cataloging in Publication Data

Sherley, Thomas, 1638-1678.
A philosophical essay.

(History of geology)
Reprint of the 1672 ed. printed for W. Cademan, London.
1. Petrogenesis. 2. Science--Early works to 1800. I. Title. II. Series.
QE431.6.P4S53 552'.03 77-6541
ISBN 0-405-10460-X

A Philoſophical ESSAY:

DECLARING
The probable Cauſes, whence Stones are produced in the Greater World.

From which occaſion is taken to ſearch into the Origin of all Bodies, diſcovering them to proceed from *water*, and *Seeds*.

Being a *Prodromus* to a Medicinal Tract concerning the *Cauſes*, and *Cure* of the Stone in the Kidneys, and Bladders of Men.

WRITTEN
By Dr. *Thomas Sherley*, Phyſitian in Ordinary to His MAJESTY.

LONDON,
Printed for *William Cademan*, at the Pope's Head, in the Lower Walk of the *New-Exchange*. 1672.

To the Illuſtrious,

GEORGE,

Duke, Marquis, and Earl of *Buckingham*; Earl of *Coventry*, Viſcount *Villiers*, Baron *Whaddon* of *Whaddon*, Lord Roſſ of *Hamlock*, *Belvoir*, and *Trusbut*, &c. Maſter of the Horſe, Knight of the moſt Noble Order of the Garter, Chancellor of the Univerſity of *Cambridge*, and one o His Majeſties moſt Honourable Privy Council.

May it pleaſe your Grace,

T*Is not the ſublime condition in which you are, nor the*

emi-

eminent, and great Honours with which you deſervedly ſhine,(as a bright Star, of the firſt Magnitude, in our little World,) that hath induced me to addreſs this enſuing Diſcourſe to you; but the great, and excellent knowledge of Natural Beings your Grace hath acquired by a conſtant, and curious Anatomizing of all ſort of Concrets in your Laboratory; a way certainly the moſt likely to give you a faithful and ſolid account of the Nature

ture of things, by diſcovering to you the real principles of which they are conſtituted. This it is, which made me conclude, I ſhould have done a great injuſtice, had I put this Tract under any other Protection than yours. And indeed, at whoſe feet can a Subject of this Nature be ſo fitly plac'd as at your Grace's, you being ſo great an Experimental Philoſopher?

But leſt I prove tedious, I will conclude this Epiſtle, with aſſuring you, that not only this Book, but the Author of it, are both Dedicated to your Graces Service, by him that in all Humility subscribes himself,

My Lord,

Your Graces

most Obedient, and

Faithful Servant,

THO. SHERLEY.

TO THE READER.

READER,

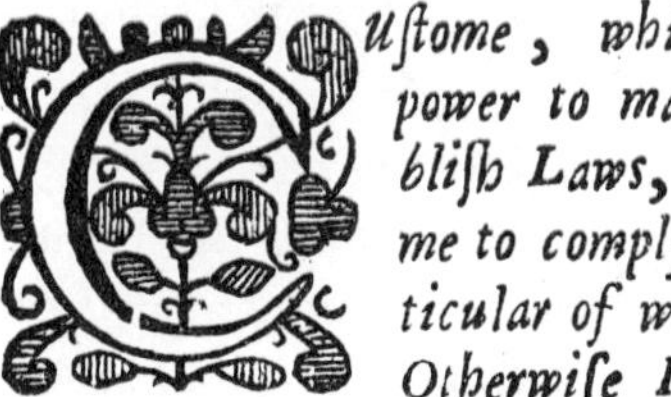

Uſtome, which hath the power to make, and eſtabliſh Laws, *hath obliged me to comply in this particular of writing to thee. Otherwiſe I was reſolved to ſuffer this enſuing* Diſcourſe *to appear naked, and without an* Advocate, [*as* Philoſophical Subjects *ought to do:*] *that ſo the minds of the ſtudious, being free from prepoſſeſſion, might be the better able to judge of the truth of the* Matter *in hand, and of the validity of the* Arguments *I produce to evince it.*

This, I ſay, I would have done, could I have been aſſur'd, that this Book *ſhould have fallen under the cenſure of none but*

Philosophical, and knowing Men, to whom I should have thought my self happy to have submitted my labours in this kind. To which sort of inquisitive, and industrious Men, I pretend not to have done any further service in these Lucubrations; then by having laid together those Arguments, and Experiments, which did readily occurr to my mind; and which I thought might conduce to prove the Matter in hand, a Subject fit to be seriously look'd into; and though I seem in some places to be determinate, yet I declare [once for all] I have not the vanity to think I have put such a Ne plus ultra *to the inquiries into this Subject, that no further discoveries are to be made; nothing less. For though the Subject be rough, and hard, yet it is far from being unfruitful. And if by my endeavours I shall prove Instrumental, [by giving of hints, &c.] to put other industrious Philosophers, who are fitted with better parts, and more time, to digg deeper in these Quarries, I shall think it glory sufficient, to have been thus far serviceable to the Common-wealth of Learning: and if by the endeavours of such Worthy Men, I shall find my self confirm'd in my Opinion, I shall rely upon it with the greater security. But if by their inquiries, other, and truer causes shall appear;*

appear ; I ſhall not ſcruple to acknowledge, that I will willingly become a Proſelyte to Truth, though at the ſame time it is diſcover'd it convince me of having been erroneous in my Opinion.

But at preſent, thinking I defend a verity, I ſhall not eaſily recede from my Opinion, without my Judgment be convinc'd, by the ſame means I make uſe of, to Proſelyte others : that is, both by reaſon, and Experiments. And likewiſe let me add this, that I ſhall expect the ſame Candid, and civil dealing from ſuch who intend to confute me, which I have ſhew'd to thoſe whoſe Opinions I reject. For otherwiſe I ſhall conclude a railing Adverſary fitter for my ſlight, than reply ; I knowing a better uſe of my time, then to ſpend it ſo unfruitfully.

As I court not applauſe, which is a vanity unbefitting a Philoſopher ; So, having [as I ſuppoſe] appear'd in a good Cauſe, that is, the defence of a Philoſophical truth, [viz. *that the Matter of Stones, and all other Bodies, is Water, and their Efficient Seed*] *I ſhall not fear Cenſure, though I muſt be expoſed to that of any Man, which ſhall take the pains to peruſe my Book ; I am not ignorant of the Proverb,* So many Men, ſo many Minds : *Nor of that other,*

other, Habent ſua fata Libelli: *And therefore cannot expect that impoſſibility of pleaſing every body; but that I may be as uſeful as I can to thoſe Readers, which though they may have large Souls, have yet been little Converſant with things of this Nature; I ſay, that I may be as Inſtructive as I can, and that my meaning may not be miſtaken, I ſhall therefore inform them of theſe things following.*

First, that there are many Men, of great Natural parts, which yet want the advantage of underſtanding the Greek, *and* Latine *Tongues; for whoſe ſakes, I have [that I might be the more uſeful] Tranſlated into the* Engliſh, *all thoſe quotations which I make uſe of, from Authors which have writ in thoſe Learned Languages; and that [for the moſt part]* Verbatim, *[though ſometimes I only deliver their Sence.] And to ſatisfie the ſcrupulous, yet Learned ſort of Readers, of my integrity, I have almoſt conſtantly given them the very words, and in the ſame Language they are delivered by thoſe I quote; together with the Book, and for the moſt part, Page, where the*

Ori-

Original words may be found, [marked in the Margin.]

Secondly, If it ſhall be objected, that I am very frequent in quotations, [a thing much out of faſhion;] and that therefore it may be ſuppoſed, I have ſaid little but what will be found expreſſed by others; I ſhall acknowledge I have wilfully done ſo, becauſe I had a deſire to get my ſelf ſtrongly Seconded in my Opinion by the determinations of Learned Men; [And of the Teſtimony of ſuch only have I made uſe.] For I verily believe, that if an Angel himſelf ſhould avouch any thing ſingly, and as his own Opinion, he would not be believed by ſome Men. But however the Reader will have theſe Advantages by it. Firſt, thoſe things are here contracted, and brought under their proper heads, which are diſperſed in many Voluminous Authors; which will ſave him time in ſearching many Books. Secondly, He may find the Pith, and Subſtance of what others have written in their Languages, delivered in his own. And thirdly, here are, beſides, many Experiments, and Obſervations of my own, very conducible [I ſuppoſe,] to clear,

and

and explicate those Philosophical Principles I have undertook to defend in this Discourse.

Thirdly, If any Man shall be so much a Momus, *as to repine at the just commendations I often give to* Van Helmont, *and Mr.* Boyl; *I must needs say, that I think his ill Nature proceeds from his want of throughly knowing these Authors: for if he had taken the pains to search the depth of these two, as I have done, I doubt not but he would acknowledge, I have fall'n short of giving them their deserved praise, [they having merited so much from all inquisitive, and Learned Men.]*

Lastly, I think it necessary to tell thee, how I would have to be understood those two words of Seed, *and* Water, *the Principles upon which I have built this Discourse.*

First then, by Seed *I understand a fine, subtile Substance, [imperciptible by our gross Organs of Sensation;] in which God hath impressed a Character of that thing he will have it produce from the Matter it is to work upon: which it doth*

doth perform by putting the parts of Matter into such a peculiar motion as is requisite to produce the intended Effect. And this we may illustrate thus.

A Woman with Child, by a strong desire, forms in her Spirits an Idea *of some Fruit she longs for; and by the powerfull motion of that* Idea *working on the Child, she forms a real Effigies of the said Fruit upon that Member of the Child which corresponds to that of her own Body she touched with her hand; which, as Experience teacheth us, will Vegetate, grow Ripe, and Wither, according to the several mutations the Fruit it resembles undergoes. And we are told by* Esdras, *that God, before he made the World, did consider the things he intended to make; and then produced them. By which Expression, I think may well be understood, the Creation of all those Spiritual, and Seminal Beings, containing in them, not only an* Idea *of the thing to be made; but also a power to move the Matter after a peculiar manner, by which means it reduceth it to a form like it self. And as a Painter doth first conceive in his mind a Spirituall* Idea *of the Picture he intend-*

eth

eth to draw; and afterwards by peculiar Motions of his hand, which are guided by the said Idea, *he produceth a perfect Picture Corresponding with that in his mind: So likewise, by putting Matter into peculiar Motions, the Seminal* Idea *makes it self visible.*

By Water, *the Material* Principle *of all Concrets, I understand, a fluid Body, consisting of very minute parts, and variously figur'd* Atoms, *or* Corpuscules, *the Mass of it being full of pores, and therefore subject to be contracted into less room: and upon the same account it doth easily, and readily submit to those motions it is put into by Seminal Beings: from which moving of Matter all the visible, and Tangible Bodies of the World, have their result: And therefore I have, all along this ensuing Discourse, took care to explicate the* διότι *of the Origin of Bodies, by the* Mechanical Principles: *That is, by the Motion, Shape, Size, Scituation, and Connexion of the parts of Matter.*

But though this be a way commonly used, in explicating things, by the Philosophers *of our Age; yet most of them leave out the first principle of Natural Motion;* viz. *the Seminal principle, which I have taken in, to compleate my Hypothesis.*

And

And now having said thus much, I shall say this further, [*and let it not be counted a vanity*] *that I think, and hope, I have in some considerable measure made out the truth of those principles I have assumed to defend.*

It hath cost me some pains to Collect, and draw into proper Sections, the Body of this Discourse: which I have also strengthned by the Authority of the best Philosophers, and Learnedst of Men, both Ancient, and Modern. All which I here present thee with; heartily wishing all ingenious Men may see the usefulness of, and receive as much satisfaction in this Doctrine; as I do, who am a Friend to all that industriously search after the Truth, and Nature of Things.

THO. SHERLEY.

From my House, in *Newton-street*, over against *New South-hampton* Building, in *High Holborn. Jan.* 27th. 1672.

The Reader is desired to Correct as he Reads, these Errors of the Press, as likewise any other he shall find.

ERRATA.

PAge 12. in the Margin, *leg. Consensus.* p. 13. lin. 2. read Concurrere. page 15. lin. *ult. leg.* δεκτικόν. p. 33. *dele* these words [or intire.] p. 34. lin. 5. *leg. à priori. ib.* lin. 8. *leg.* Springy. p. 35 lin. 11. *dele* whilst, and they. p. 16. *dele* [. p. 38. in the Margin, *leg. Elementis.* p. 40. lin. 23. *leg.* fæces. p 103. lin. 25. *leg.* seminal. p. 126. lin. 26. *leg.* apposition. p. 124. lin. 24. *leg.* ἀληθεύειν. p. 110. lin. 28. *leg.* those. p. 137. lin. 1. *leg.* least. p. 129. lin. 1. *leg.* Ætherei. p. 114. lin. 1. *leg.* [ἰλν.] p. 119. lin. 9. *leg.* ὄντον.

THE FIRST ESSAY:

Being a Difcourfe intended to demonftrate, that not only Stones, but all other Bodies, owe their Original to Seeds, and Water.

Section the First.

HAving, in complyance with the importunate defires, or rather commands, of many of my Worthy, and ingenious Friends, obliged my felf to acquaint the World with my thoughts concerning the moft probable caufe of the Stone, both in the Kidneyes, and Bladder of Men; and having begun a Tract upon that Subject; I fore-saw a necessity, [before I suffer'd that discourse to appear in publick] to inquire into the Causes, and Nature of Petrifaction [in the greater World] in general: and I was encouraged the more to do so, by a Passage I met with in the Works of that Noble Philosopher, Mr. *Boyl*, whose words are these. *Since we know very little a* Priori, *the observation of many effects manifesting, that Nature doth actually produce them so, and so, suggests to us several wayes of explicating the same* Phænomenon, *some of which we should, perhaps, never have else dreamed of; which ought to be esteemed no small advantage to the Physitian:* And again; *He that hath not had the curiosity to inquire out, and consider the several wayes whereby Stones may be generated out of the Body, not only must be unable, satisfactorily to explicate, how they come to be produced in the Kidneys, and Bladder; but will perhaps, scarce keep himself from embracing such errors, (because Authoriz'd by the suffrage of eminent Physitians) as the knowledge I am recommending,*

Boyl, Usefull. of experiment. Philosophy. p. 31.

ing, would easily protect him from.

2 Let us then, in the first place, examine, how Nature produceth Stones without the Body of Man (that is, in the greater World;) after which we will see, if the causes of generating Stones in the Bodies of Animals, be not the same; or at least, bear some Analogy, or resemblance thereunto. Which that we may the better be enabled to do, I shall relate some choice Histories of Petrifications, taken out of approved Authors; and then examine the causes by which they were performed.

3 *Gabriel Falopius* mentioneth a River, called *Else*, which receives into it self the Torrent of the River *Sena*; into which, Wood, Herbs, or any other thing being cast, it converts it into stone. History 1. *Falopius de Metal. & fossilibus.*

4 *Albertus Magnus* relates, that in the *Danish* Sea, near *Lubeck*, in his time, there was found an Arm of a Tree, with a Nest, and Young Birds in it, the Wood, Nest, and Birds being all converted into Stone. 2.

5 *Domitius Brusonius* tells us (not upon hear-say, but upon his own knowledge) that the branches of Trees, with their Leaves, being cast into the River of *Sylar*, do turn into stone. 3.

4. *Marbodius* acquaints us, that there is a 6
De Lapide, ex Alberto. Lib. 1. Mineral. Cap. 7. Fountain in *Gothia* [or *Guthland*] that changeth whatsoever is put into it into stone; and that the Emperour *Frederick* being incredulous of the thing, did send his Glove thither, sealed with his Ring; & that that part of the Glove, with the seal, which was immersed in the Water, was in a few dayes converted into stone; the other part remaining Leather.

5. *Johannes Kentmannus*, concerning 7
De fossilibus. *Fossils*, writes, that Arms of Trees, with the Leaves, Bark, Wood; also Gloves, and divers other things, being cast into a certain Fish-pond, near the Castle of *Schellenberge*, in *Misnia*, are turned into stone.

6. *Bartholomæus à Clivola* affirms, there 8
In Lib. de Balneis. is a Lake betwixt *Cæsarea*, and *Tuana*, two Cities of *Capadocia*, into which part of a Reed, or Stick being put, it by degrees is changed into stone, that part which is out of the Water remaining what it was before.

7. *Anselmus Boethius* declareth, that in 9
Lib. 2. de Lapid. & Gem. Cap. 300. *England*, near the River *Dee*, by *West-Chester*, there is a great Cave, into which whatsoever water flows, is turned into stone.

8. *Thomas Moresinus* relates, that in 10

Moravia

Moravia there is a dark Water, in which there doth not at all appear any viſcous matter; which water, nevertheleſs, coagulates into ſtone.

11 *Johannes Petrus Faber* giveth us a wonderful account of a Spring in the Suburbs of *Claremont*, in the County of *Avernia*. *It flows* [ſayes he] *out of a Rock, and in its very coming forth it produces Rocks, and white ſtones; and the Inhabitants of this City, when they would make a Bridge to go over any of the ſmall Rivulets, which are made by this Fountain, that ſo they may viſit their Fields and Gardens, do thus: They cauſe the Water of this Fountain to glide over certain planks, made Arch-like, and within twenty four hours they have a ſolid ſtone Bridge; by the help of which they can paſs dry-foot over the Rivers. The Water of this Fountain is viſibly changed into ſtone, yet nevertheleſs it alwayes flows as other Springs do: This water is exceeding clear, nor doth it differ in colour, or clearneſs, from other Springs; Beaſts will drink of it if they be not hinder'd; but if they do, it coagulates in their ſtomacks into ſtone, from whence Death follows, by reaſon of a Collick cauſed from thence, which kills*

History 9. *In Lib. Hydrogr. Spagyr Cap.* 14.

 kills

kills with cruel torments all the Beasts that have drunk this Water. Wherefore the Inhabitants take care to drive their Cattel far enough from this Fountain; for it is as a present poyson to all sorts of living Creatures that drink of it. When it is taken from the Spring, it is quickly turned into stone; the truth of which the Inhabitants do make manifest [to all that doubt thereof] by many experiments; they fill a glass with this Water, and presently it is converted into stone, which retaineth the shape of the glass: so likewise if Earthen Vessels be filled with this water, it is suddenly congealed into stone, which keeps the form and figure of the Vessel that contained it. This wonder of Nature [sayes he] *every body admires, but I believe hardly any body will be found, that shall be able to render the Natural reason of this thing.* Thus far he.

10. *In vita Peireskii. Lib. 1.*
Gassendus tells us, that *Peireskius* [ac- 12
cording to his usual custom in the Summer] going into a stream of the River *Rhosne*, to wash himself; he observed once the ground to be hard under his feet, and uneven, [which had at all times before been soft, and smooth] being full of knobs, and Balls, about the bigness,

and

and likeness of Eggs boyled hard, and the shells pilled off; which he looking upon as somewhat strange, took some of them up, and carried them home; but a few dayes after he was surprized with a greater Admiration: for, going again into the same place of the River, he found those soft, and yielding lumps, he had left there, turned into perfect pebble stones; and also viewing those he had laid up at home, he found them likewise turned into true Pebbles.

13 *Helmont* likewise affirms, that [contrary to the Proverb, *Gutta Cavat Lapidem*, A drop by often falling doth hollow a stone] there is a Spring in the Monastery of *Zonia*, near *Brussels*, that breeds stones so fast, that the Monks are daily forced to break them off with Crooks and Hatchets. 11. *De Lithiasi. Cap.* 1.

14 And I my self have seen a Spring near *Wrixham*, in *North-wales*, that in a short space of time would convert Sticks, Straws, Leaves, Leather, or any other subject, put into it, into stone. And of this Nature are divers other Springs to be found, both in *Ireland*, and *England*. 12.

15 Our Industrious Countrey-man, *Gerard*, assureth us, he knew several Springs 13.

History of Plants. Lib. 3. p. 1586. of this Nature, both in *England*, and *Wales*: As in *Bedford-shire*, in *Warwick-shire*, near *Newnam Regis*; and another near *Knasborrow*, in *York-shire*; he likewise tells us, he knew divers pieces of Ground, into which a stake being struck, that part in the ground would be changed into stone, the other part remaining Wood.

14. In apend. Syntag. Arcan. Chym. Cap. 32. *Libavius* relates, That a certain 16
Hen sitting on her Eggs, being struck with a Gorgonick Spirit, was transformed into stone, with her Eggs likewise.

15. In Præfat. Lib. de signat. Rerum. *Crollius* relates, that in a certain place 17
of *Moravia* there is a stupendious Den, in which are to be found divers, and admirable sportive works of Nature: for the drops distilling from the upper part of the Cave, into the hollow of it, do there form many intricate Labyrinths in the Mountain, and do presently [of their own accord] convert into stone, by the help [as he thinks] of the Spirit of Salt; and in their falling from on high, they form various Figures, and Statues of stone.

16. *Aristotle* sayes, that in the Metalline 18
Grots of *Lydia*, about the City *Pergamos*, certain Workmen, in the time of War, having fled into them to hide them-

themselves, and the mouth of the Cave being stopp'd, they perished there; but afterwards being found, not only their Bones, but their Veins, with the humours contained in them, were found to be turned into stone.

19 *Aventinus* also writes thus: In the 17.
Year 1348. by an Earthquake, more *In Histor.*
than fifty Country men, with their Milch Bavar.
Lib. 7. id
Cows, and Calves, being killed and *est. in Anal.*
stifled by an Earthy saline Spirit [as he Bavar.
supposeth] they were reduced into saline Statues, [such as *Lots* Wife:] And this happened amongst the *Carini* [a People of *Germany*;] which similitudes or Images of Men, and Beasts, were seen both by him, and the Chancellor of *Austria*.

20 To the like purpose, *Helmont* tells us 18. *De Lithiasi*
of a whole Army, consisting of Men, *Cap.* 1.
Women, Camels, Horses, Doggs, with their Armour, Weapons, and Waggons, which were all transmuted into stone, and remain so to this day, [a horrible spectacle;] And this, saith he, happened in the Year 1320. betwixt *Russia* and *Tartary*, in the Latitude of 64. degrees, not far from a Fen of *Kataya*, a Village, or Horde, of the *Biscardians*; which he very rationally con-

concludes to have happened from a strong hory petrifying breath or Ferment, making an eruption through some clefts of the Earth, the Land being stony underneath; and the Winds having been silent for many dayes.

He that desireth more Examples 21
of this kind, let him consult Gorgius Wernerus, *de Ungaricis.* Godfrid. Smoll. *in lib. Princip. Philosoph. Et Medic. antiquitatis. Cap.* 10. F. Leander Albertus *in descript. Italiæ.* Andreas Laurentius, *lib.* 2. *de strumis. Cap.* 2. Georgius Agricola, *lib.* 7. *de Natura fossil. Cap.* 22. Johannes Wigandus, *in libello de Succino*: Lobelius, *in fine Observat.* Cælius, *&c.* But I suppose what I have here related sufficient; and therefore I think it now time to inquire into the Causes of Petrification, and the Efficients of these Transmutations.

Section the Second.

22 THe Doctrine of the four Elements [with their qualities, concurring, as is suppos'd, to the production of Bodies, which was introduced by the Authority of *Aristotle*, and hath since prevailed with most Men even to this Age of ours,] hath been the cause, why we have hitherto received but an unsatisfactory account, not only of the Origine of all concretes, but more particularly concerning stones; and that not only in Relation to the Material Cause, but also to the Efficient, of Petrifications in general.

23 For, they seem to think it sufficient, to have crudely told us, that Stones [and all other Minerals, and Metals] are made of Earth, with a slight mixture of the other three Elements, as the Material; and by the assistance of Heat, Cold, Moisture, and Driness, as the External, and efficient Cause. For perceiving the weight of Minerals, and Stones, to exceed the weight of water, they therefore assign the matter of Minerals, and

and Stones, to be chiefly Earth; and without any further Controversie, or search after the matter, they are content to believe, and would have us do so too, that all sorts of stones are nothing but Earth, from which the other three Elements are forced by heat; by which means it becomes baked into a stone. And this they [viz. the *Aristotelians*] think they prove by alleadging the Example of Potters Earth, which being burnt gains a stone-like hardness. And because neither Stones nor Earth do commonly melt in the fire, they therefore conclude stones are made of Earth. But there being no such heat in the Superficies of the Globe, much less in the bottom of the Water [where commonly stones are bred,] I must confess I can receive but little satisfaction from this account.

And I find the Learned *Sennertus* is as 24
unsatisfied with this Doctrine as my self: for he will by no means allow the Elements, or their qualities, to be the Primary Efficients of Stonification. His words are these; *Licèt vulgò multi é qualitatibus primis Calculorum Concretionum & Coagulationum causas deducere conantur; tamen frustra laborant. Nam neque exsiccatio, nec calor, nec frigus, hîc locum habere possunt,*

Sennertus, *in Lib. de Consens. Chymic. Cum Galenist. Cap.* 2.

possunt, ut primariæ causæ, [nam, ut causam sine qua non, concurre posse, non negamus; dum scilicet aquam, quæ concretioni obstat, absumit;] neque à quoquam hactenus commonstrari potuit, quomodo calor nudus talem Concrescendi dispositionem generare, & succum Lapidescentem producere possit. Imo fit hoc etiam, ubi omnis Calor abest, & in frigidis etiam membraneisque locis, item & in Infantibus, ubi nullus concedatur Caloris excessus, sed manifesta potius cruditatis indicia deprehendantur, in vesica generantur Calculi; & quomodo, quæso, in fontibus frigidis, in quibus ligna immersa in lapides transformantur, succus lapidescens à Calore producitur? Deinde, frigus quod attinet, non semper in loco frigido, vel minus calido, Calculi concrescunt, cùm & in capite, & in pulmonibus, circa basin Arteriæ magnæ, in Cordis arteriis, imo in Corde reperti sint: Uti Legimus in Observation. Cornel. Gemmæ, *lib.* 1. *Cosmocritic. Cap.* 6. Anton. Beniven.. *de abdit. Morb. & Sanat. Causs. Cap.* 24. Fernel. 5. *Patholog. Cap.* 12. Hollerii, 1. *de Morb. internis, in schol. Cap.* 29. & 50. *Et in balneis etiam Calidissimis Trophos ac stirias saxeas concrescere, ubi frigus nullo modo admitti potest, experientia compertum*

compertum habetur: in *English* thus;

"Though it hath been much endea-
"vour'd by many, to deduce the causes of
"the concretion, & coagulation of stones,
"from the first, or primary qualities, yet
"hath their labour been in vain: for nei-
"ther can drought, heat, or cold, be here
"allowed as a primary cause, [but we do
"not deny, that they may concur as a
"cause, *sine qua non*, so that it may, for
"Example, waste the water, which hin-
"ders concretion;] neither could it hi-
"therto be demonstrated by any body,
"how heat of it self could be able to ge-
"nerate such a disposition of compaction;
"and that it could produce a Lapidescent
"juice: Nay, this is performed where all
"heat is wanting, and that in cold and
"Membranous places; as also in Infants,
"who are not allow'd to have any excess
"of heat, but are rather found to have
"manifest crudity, the stone is genera-
"ted in the Bladder: and how, I pray, is
"the stonifying juice produced in cold
"Fountains, into which wood being cast
"is changed into stone? Then, as to cold,
"stones do grow in the Head, in the
"lungs, about the basis of the great artery,
"in the Arteries of the Heart; nay, they
"are

"are in the Heart it ſelf. Alſo there grows "in hot Baths, as experience ſheweth, "ſandy ſtones, & ſtony Iſicles, where cold "can by no means be admitted. Thus far he: by which you ſee he is clearly of opinion, that neither heat, nor cold can be the primary, or chief cauſe of Petrification; contrary to the Axiom which *Ariſtotle* layes down, to this effect;

26 *Of thoſe bodies which adhere together, and are hard, they are wont to be thus affected; ſome by the fervour of heat, ſome by cold; that drying up the moyſture, this preſſing it forth.* *In Meteorologicor. Lib. 4. Cap. 8.*

27 Let us then inquire what the Chymical Philoſopher's opinion is in this point: (and the rather becauſe it is conſtantly affirmed by moſt of them, that the Art of Pyrotechny is the only true means of informing the mind with Truth, and acquainting it with realities; and we ſhall find, that they hold Salt to be the principle of ſolidity, and the genuine cauſe of coagulation, in all bodies; [as alſo of ſtonification:] For, ſay they, if you conſult experience, all thoſe things that are compact, or ſolid, do contain Salt; and where there is no Salt, there can be no hardneſs. And for this reaſon they eſteem Salt to be the πρῶτον δεκτικὸν of Soli-

Solidity: which they that deny [say they] are obliged to shew some other cause; from which Salts have that aptitude to coagulate themselves, and become solid bodies.

For, it is manifest, that the Salts of 28
Vegetables, as Crystals of Tartar, &c. also Nitre, Allom, Vitriol, Salt Gemm, [and divers other of this Nature] do coagulate themselves, not only into hard, but even brittle bodies, in the bosome of the water; and to this end they alleadge, that if the Salt be washed from ashes, no heat of fire will make them hard; but if the Salt be left in them, [and they be mixt with a little water] the fire will not only quickly make them become hard; but if they be strongly press'd with it, turn them into Glass.

Kircherus *in Mund. Subter.*

The Learned *Kircherus* is also of the 29
same opinion with the Chymists, [*viz.* that Salt is the cause of stonifying] and giveth us this experiment to confirm it. *Si saxum* [inquit] *quodcunq; in tenuissimum pollinem resolveris, & aqua perfectè commixtum, per Manicam Hippocratis Colaveris, illa nil prorsus saxeum, sed præter arenaceum solummodo sedimentum nil relinquet; si verò Nitrum, vel Tartarum, aqua perfectè commixtum addideris,*

illa,

illa quæcunq; tetigerint intra subjectam concham posita, sive frondes, similiaque, post exiguum temporis curriculum aeri exposita, vel in saxum ejusdem generis conversum si non totum, saltem cortice Saxeo vestient. “*If* [saith he] *you reduce any sort of stone into a most subtile powder, and mixing it throughly with water, you strain it through* Hippocrates's *bagg, there will nothing of it remain that is stony; nor will it leave any thing of it behind, but a certain sandy sediment; but if you shall add to this, Nitre, or Tartar, perfectly dissolved in water, whatsoever body they shall touch, being placed in the same Dish, whether it be the twiggs of a Vine, or the like, after a little while being exposed to the Air, it will be turned into stone; or at least it will be covered with a stony Crust.* And though this opinion be held by *Crollius, Hartman, Quercetanus, Severinus*, and *Sennertus*, [who are but Neoterick, or late Writers] yet is it no new opinion, but hath been asserted by the venerable Ancients, as long agoe as the time of *Hermes Tresmegistus*, [who is said to have lived in the Age of *Joshua*] who in his *Smaragdine* Tables [as they are called] hath left us these words. *Salis est, ut corporibus in Mundum prodituris, soliditatem coagulando præstet; Sal enim corpus est, Mercurius Spiritus, Sulphur anima,* that is; “ *'Tis from Salt that Bodies are produced in the World; it causeth Coagulation, and Solidity: for Salt is the Body, Mercury the Spirit, and Sulphur the Soul.*

This Doctrine, though much more 31
rational than the former, and seeming to be confirmed by experiment, and to be verified by the account our senses give us of it, cannot yet gain my full assent to it, so far as to allow Salt to be the Primary, either Matter, or Efficient of Solidity in bodies, or the cause from whence stones are produced. For it is observable, that Salts are reducible into Liquors, [and do seem to lose their solidity] either by being mixed with water, or exposed to the Air, in which many of them run *per deliquium.* But, to let this pass; what Salt can be supposed to be communicated to Quick-silver, when it is coagulated by the fumes of melted Lead, by which it becomes so solid, that it may be cast into Moulds, and Images formed of it; and when cold, is not only hard, but somewhat brittle, like *Regulus* of *Antimony*? What access of

Salt can be fancied is added to the white of an Egg, [from whence the whole Chick is formed] which is a Liquor so near water, that by beating it with a whisk it is reduced into so fluid a substance, that it will easily mix with water, and is hardly distinguishable from it? And yet this white of the Egg, by the assistance of a gentle heat, to stir up its seminal Principle, and enable it to turn, and new shuffle the parts of that liquid substance, [by the means of which motion divers of its parts are broken into shapes and sizes fit to adhere one to another] is all of it turned into solid bodies, some of them very tough, as the Membranes, and Nerves; and some of them hard and brittle, as the Beak, Bones, Claws, &c. [of the Chick;] and all this without any new addition of salt.

'Tis likewise remarkable, that very credible witnesses assure us, that Corral [though it grow in salt water, at the bottom of the Sea] is yet, whilst it remains there, soft, like other Plants; [and juicy also:] neither will the example of *Kircherus*, alleadged above, avail much; since it is commonly known, that the powder of Plaster of *Paris*, or burnt Alabaster, if it be mixed with water,

Gassendus, Lib. 4. *Anno Dom.* 1624. Mr. *Boyl.* Essay of firmness.

water, without any ſort of ſalt, will coagulate into an entire ſtony lump, or Maſs.

I do not deny but that ſalt may very much conduce towards the coagulation of ſome bodies, as we ſee in the curdling of Milk with Runnet, Spirit of ſalt, Oyl of Vitriol, juice of Limmons, and the like; but then this happens but to ſome bodies, and is cauſed from the ſhape and motion of its ſmall parts, which entring the pores of ſome bodies that are naturally fitted to be wrought upon by it, it fills up many of the cavities of ſuch bodies; and alſo affixing it ſelf to the particles of them, it cauſeth them, not only to ſtick to it ſelf, but alſo adhere cloſely one to another. 32

I ſay, ſalts do this to ſome bodies [not to all,] for to ſome other bodies, inſtead of being an Inſtrument, either to cauſe, or confirm their ſolidity, it by diſſociating the parts, of which they conſiſt, and putting them into motion, doth reduce them into the appearance of Liquor; as we ſee in the action of corroſive ſaline ſpirits, both upon Metals, and ſtones. 33

Now, for that Argument, that ſalts do ſhoot even in the water into hard, and 34

brittle

brittle Cryſtals, if I ſhould ſay they do ſo upon the account of a ſeminal Principle, I ſhould not, perhaps, be thought to have much miſtaken the cauſe, by thoſe that have well conſider'd the curious and regular Figures [yet conſtantly diſtinct from each other,] which their Cryſtals ſhoot into : which certainly cannot proceed from chance ; for they do as conſtantly keep their own figure [as for Example, that of Nitre alwayes appears in a Sexangular form, that of Sea-ſalt in a Cubical :] As Wheat produceth Wheat, and the ſeed of a Man, a Man.

35 Philoſophers hold, there are two ſorts of Agents ; one they ſtile αἴτιον, that is, the principal cauſe, or Agent ; from which immediately, and primarily, the Action depends, and by whoſe power the thing is made ; and this [as we ſhall prove in its due place] is an Architectonick ſtonifying Spirit, or Petrifick ſeed. The other cauſe they call συναίτιον, or the Adjuvant, or aſſiſting cauſe, [of which ſort there are many] by which the principal Agent may be furthered in its acting upon matter ; of which laſt ſort of cauſes [of the ſolidity in Bodies, *viz.* the Helping, or Aſſiſtant] we

will not deny but that salt may be one, as being such a prævious disposition of the parts of Matter, as renders them more apt to be wrought upon by the first kind of Agent, *viz.* the Seed. So that in some sence we may [for the reasons above alleadg'd] allow the Chymist to think salt is [though *Nec prima materia, nec efficiens.* Yet] *Proxima materia, &* συναιτιον *Soliditatis.* "*The Proximate* "*matter, and Adjuvant cause of Solidity.*

But since not only salt, but the whole 36
Tria prima, or Three first Principles of the Chymists, as also the *Quaternary*, or four Elements of the Peripateticks, are justly enough denyed to be the first Elements, or constitutive Principles of all Bodies, [they themselves being further resolvable into more simple parts, as we shall prove by and by.] I say, since it is so, I must be excused, if denying my suffrage to both their Doctrines, [in that large sence they propose it in :] I offer to render other causes, by which not only solidity, but Petrification also may be introduced into Matter.

Section the Third.

37 THe Doctrine I am now about to affirm, is no Novel conceit; but so Ancient, that we shall find that it was held, [and by them transmitted to Posterity] not only by *Plato*, *Timæus Locrus*, *Parmenides*, *Pythagoras*, *&c.* Philosophers of the Academick, and Italick Sect; but also by *Orpheus*, *Thales* the *Milesian*, and also by *Mochos*, and *Sanchoniathon*, the great, and Ancient *Phænitian* Philosophers; nay, by that Divinely illuminated Man, *Moses*.

38 I urge this point of the Antiquity of the Doctrine I am now going to affirm, because I know it is the custom of some Men, to disgust any Philosophical truth, that cannot shew it self to be as ancient as *Aristotle*'s time; but to please such, let them consider, that the *Hypothesis* we intend to make use of in this ensuing Discourse, beareth an equal Date with the World, and was at first deliver'd to Man by the *Ancient of Dayes* himself.

This Doctrine then [which hath of

late years been revived, and assumed by the Noble *Helmont*, and other great wits,] I now am come to lay down, and explain; and in the next place shall endeavour to prove, and confirm it; first, by reason, then by experiment, and lastly, by Authority.

The *Hypothesis* is this, *viz.* That 40
stones, and all other sublunary bodies, are made of water, condensed by the power of seeds, which with the assistance of their fermentive Odours, perform these Transmutations upon Matter.

That is, that the matter of all Bodies 41
is originally meer water; which by the power of proper seeds is coagulated, condensed, and brought into various forms, and that these seeds of things do work upon the particles of water, and alter both their texture, and figure; as also, that this action ceaseth not, till the seed hath formed it self a Body, exactly corresponding with the proper Idea, or Picture contained in it. And that the true seeds of all things are invisible Beings, [though not incorporial;] this I affirm, and shall endeavour to prove.

But to make this the better to be un- 42
derstood, I shall præmise some generals, and

and then deſcend to particular proofs of what I aſſert.

43 Firſt then, nothing is produced by chance, or accident. And therefore in every Generation, or Production, there muſt neceſſarily be preſuppoſed ſome kind of ſeed which hath a power, or faculty, to alter the Matter, and diſpoſe it to ſuch a Being, and Form, as God and Nature have deſign'd to produce.

44 Secondly, all ſeeds (in ſome degree) are endow'd with Life, and a power of acting: for nothing that is not Vital can promote it ſelf to perfection. And if Bodies are diſtinguiſhable from their internal Efficients, and are ſpecificated by them, then muſt they be allowed to contain a ſeed.

45 Theſe poſitions will not [perhaps] be denyed to Animals, nor Vegetables; becauſe their ſuppoſed ſeed is viſible. For the ſeed [or rather, ſperm] of perfect Animals, is an effloreſcence of the beſt parts of the blood [elaborated in the Teſticles] and impregnated with Spirits from all parts of the Body; in which reſideth the *vis Plaſtica*, or *Efficient*; [and this indeed is the real ſeed, or geniture, though it be inviſible] which containeth in it ſelf the Image, or Type of

of the thing to be made; which it performs by a Fermental *Odor*, or *Aura*, and by breathing upon those proper juices it finds in a Female Womb; it first coagulates them, and then by degrees explicates it self, working this Female Matter into a Body exactly corresponding with its own pre-conceived Figure: the grosse body of the Male-seed all this while being but a vehicle, to convey with safety this subtile fermentative breath to its proper place of action; which being done, the body of the sperm is ejected from the Womb, as useless to Generation.

That this is so, hath been proved by 46 the industrious and curious dissection of divers sorts of Beasts, made at several seasons after their Conceptions; and continued till the formation of the *fœtus*; and yet no *Vestigiæ*, or foot-steps of the Male-sperm could be found in the womb. This is asserted by that incomparable Man, Dr. *Harvy*; to whom I refer him that desireth further satisfaction in this point.

Dr. Harvy. de generat. ex Ovo.

The sperm of Man, if but for a mo- 47 ment it be exposed to the touch of the external Air, becomes dead, and unprolifick; and that by reason of the subtilty of of the ſpermatick ferment, [it being very apt to deſert the body of the ſeed.] This is a truth ſo generally known, that the Virtue of that Lady is juſtly ſuſpected by all rational Men, who pretended to have Conceived with Child, by attracting the ſeed of a Man which floated in a Bath, wherein ſhe Bathed her ſelf.

48 As to Vegetables; They alſo take their beginnings, are propagated, and do fructifie, from the like inviſible cauſe; *viz.* a fermentative Odor, [or Aura] which alſo contains the *Idea* of the Plant to be produced.

The body of the Seed, or Grain [which is the Casket that contains this inviſible Workman] being committed to the Earth [its proper Womb] is ſoftened by the Nitroſulphurous juice of the ſoyl; that the *Vis Plaſtica* [which is the Efficient of the Plant] may, being looſened from its body, be at Liberty to act. Which being done, the body of the ſeed, or Grain, is deſtroyed; according to the ſacred Writ: [*Except Seed, committed to the ground, dye, it produceth no fruit* :] But the Architectonick Spirit being now at Liberty, ferments, by its Odor, the Liquors it finds in the Earth, converting them

them into a juice, fit to work the Plant out of it, which it by degrees performs. [This Liquor in the Earth, is by *Paracelsus*, and *Helmont*, by a Barbarous name, call'd *Leffas Terræ*; and is the proximate matter of all Vegetables.] For proof of what I seem to have with some boldness asserted in this place; Let any sort of Grain be put for a small time in an Oven, [or any analogous heat,] that the external warmth may suscitate and excite this ferment of the Seed to take wing, and desert its body; This Grain, though entire to sight, if it be committed to the Earth, shall never by any Art be brought to produce its like.

As Vegetables, and Animals have their 49
Original from an invisible Seminal Spirit, or breath; so also have Minerals, Metals, and Stones.

Dr. *Jordan* of Natural Bathes. Cap. 2. p. 58, 59.

To this purpose Dr. *Jordan* tells us, 50
There is a Seminal Spirit of all Minerals in the Bowels of the Earth, which meeting with convenient Matter, [what that is, we shall shew in its place] *and Adjuvant Causes, is not idle, but doth proceed to produce Minerals, according to the Nature of it, and the Matter which it meets withal; which matter it works upon as a Ferment, and by its motion*

procureth an actual heat, as an Instrument to further its work; which actual heat is increased by the fermentation of the Matter.

51 *The like we see in making of Malt, where the Grains of Barley being moyst'ned with water, the Generative Spirit in them is dilated, and put in Action; and the superfluity of the water being removed [which might choak it] and the Barley laid up in heaps, the Seeds gather heat, which is increased by the contiguity of many Grains lying one upon another. In this work Natures intent is to produce more individuals, according to the Nature of the Seed; and therefore it shoots forth in spires; but the Artist abuses the intention of Nature, and converts it to his ends, that is, to increase the Spirit of his Malt.*

52 *The like we find in Mineral Substances, where this Spirit, or Ferment, is resident, as in Allom, and Copperas-Mines; which being broken, exposed, and Moystned, will gather an actual heat, and produce much more of these Minerals than else the Mine would yield; as* Agricola, *and* Thurniser *do affirm, and is proved by common experience. The like is generally observed in Mines, as*

Agricola,

Agricola, Erastus, Libavius, *&c. do avouch out of the daily experience of Mineral Men; who affirm, that in most places they find their Mines so hot, as they can hardly touch them; although it is likely, that where they work for perfect Minerals, the heat, which was in fermentation whilst they were yet in breeding, is now much abated, the Minerals being now grown to their perfection. And for this heat we need not call for the help of the Sun, which a little Cloud will take away from us; much more the body of the Earth, and Rocks; nor for subterranean fires. This imbred heat is sufficient, as may appear: also by the Mines of Tinglass, which being digged, and laid in the moyst Air, will become very hot; so Antimony and Sublimate being mixed together, will grow so hot as that they are not to be touched. If this be so in little quantities, it is likely to be much more in great quantities, and huge Rocks. Heat of it self differs not in kind, but only in degree, and therefore is inclined no more to one Species, than to another; but as it doth attend, and serve a more worthy Superiour, such as this Generative Spirit is.* Thus far he.

Moreover,

Moreover, that Minerals, and Metals have their proper Seeds, hear further
53 what a Myſtical Chymiſt, (but a very rational Man) *Coſmopolita* ſayes, *Semen Minerale, vel Metallorum, creat natura in viſceribus terræ; propterea non creditur tale ſemen eſſe in rerum natura, quia inviſible eſt.* "*Nature doth Create the* "*Mineral, or Metalline Seed, in the Bowels* "*of the Earth; therefore it is not believed,* "*that there is ſuch a Seed in Nature, be-* "*cauſe it is inviſible.* And the ſame Author again, thus: *Et quam prærogativam vegetabilia præ Metallis habent, ut Deus illis ſemen inderet, & hæc immeritò excluderet? Nonne ejuſdem dignitatis Metalla apud Deum, cujus & arbores? Hoc pro certo ſtatuatur, nihil ſine ſemine creſcere; ubi enim nullum eſt ſemen, res eſt Mortua;* that is, "*And what prerogative have Vege-* "*tables above Metals, that God ſhould* "*put Seed into them, and undeſervedly* "*exclude theſe? Are not Metals of the* "*ſame dignity with God that Trees are?* "*This may be held for certain, that no-* "*thing doth increaſe without Seed: for* "*where there is no Seed, that thing is* "*Dead.*

Nov. Lum. Chym. Tract. 6. p. 319.

So that it is plain, you ſee, by the afore-cited Authorities, that Minerals,

and

and Metals have Seed, & that this Seed is invisible; and that it works by the help of its ferment, or as a ferment. That stones grow, common experience teacheth us; as also the tenth History alleadged, in the first Section of this present Essay; and consequently must be endowed with seed, and ferment; so that here is, at least, an analogous way of production to that of Animals, and Vegetables (which we have declared above) and was the thing we intended here to prove.

But before I proceed, that I may be 55
the more clearly understood, I shall declare what I understand by the Ferment of the seed. The word *Fermentum*, which signifieth Leaven, is by some esteem'd to be *quasi fervimentum*, or a thing made hot; and generally is used to denote, not only a turgescence, and dilatation of the parts of Matter, (as in Leavened Bread, &c.) but also signifieth the working of any sort of Liquor, till it become Maturated, and exalted into a generous, and sprightly Drink. Fermentation is thus defined by the Learned Dr. *Willis*: *Fermentatio est motus intestinus particularum, seu principiorum cujusvis corporis, cum tendentia ad per-*

fectionem

fectionem ejusdem Corporis, vel propter mutationem in aliud; "*Fermentation is an intestine [or intire] motion of the Principles, or particles, of which any Body consists, with an intent to perfect the said Body, or change it into another.* Ferments then are subtile, tenuous Bodies, [which we generally call Spirits; for as to Leaven, Yeast, &c. they are but the cloathings of these Spiritual, and finer Substances; as we before shewed the Grains of Vegetables, and the Sperm of Animals were:] which fine subtile breath (the Ferment) hath an expansive power; by which, being immersed in any Matter, or Substance, it desiring to dilate it self, variously agitates the small particles of that matter it is joyned to, and making Excursions through all parts of the Subject it is resident in, it adhering intimately to every small patt of the Matter, doth first by the peculiar motion it hath put them into, alter and break the particles into new shapes, and sizes; and then by conveneing together with them, constitute a new texture of that Matter; and thus a new Concrete is made by the power of the Ferment.

So that, in truth, the Ferment of a

Seed, [I mean Natural Ferment] is not any Substance distinct, or separable from the Seed it self; since it is connatural 56 with it, and intimately the same, [and is indemonstrable *à priore*, as well as the Seed, and may be thus defined.

A Ferment is an Expansive, Elastick, 57 or Springy power of the Seed of any thing; by which motion of its self it also moveth the smallest particles of that Matter in which it is immersed: by which motion also [which is of divers kinds, according to the variety of Seeds] the particles of Matter acquire new shapes, sizes, and postures amongst themselves; and so a new texture of the whole is produced, agreeable to the peculiar Nature of the Seed, and correspondent to its Idea; [which Idea we shall explain in its place.]

We have likewise declared often, that 58 seeds do operate by Odors, or scents; which we think is not said without cause: for if it be well observed, it will be found, that no seeds do generate; but in the time of their acting upon the Matter there are specifick Odors produced; that is, while they are in Fermentation, and the work incompleat: for, when the Concrete is perfected, the Odor is much

much abated: [as, not to inſtance in artificial things, making of Malt, the fermenting of Beer, and Wine, in the Barrel, and the leavening of Dough, &c.] for 'tis obſervable, that the Grains of Wheat, or other Vegetables, ſown in the ground, when their inviſible ſeed begins to ferment, do ſend forth Odors; ſo alſo the Eggs of Birds, on which the Hen hath ſat. And that Minerals, and Metals, whilſt in their making they do ſend forth ſuch plenty of ſtinking Odors, that many times the workmen in Mines are ſuffocated therewith, no body can be ignorant. Now theſe Odors are fine and ſubtile *Effluviums*, [or ſmall particles of the Matter now put into motion by the power of the ſeed, Ferment: which having extricated themſelves from their Companions, and roving in the Air, do at laſt ſtrike againſt thoſe parts of our Noſes that are fitted by Nature to be ſenſible of the touch of ſuch very ſmall Bodies.

59 Odors then are a ſign of Fermentation begun, and are nothing but ſmall particles of Matter got looſe from their Fellows, begun to be alter'd, and ſpecificated by the ſeed; and therefore are very various, according to the diverſity of

seeds, and their Ferments, from whence they proceed.

Having before declared, that all Bodies proceed, and are made from Seminal Beings; and that the real seeds, and Ferments of things are invisible; and having declared, what I would have understood by a seedy Fermenr, and Odor; and also having hinted above, that all Bodies are Materially [and Primarily] nothing but water; I shall now endeavour to prove the same more fully, and clearly; the which I shall do by three sorts of Arguments. The first is grounded upon tha· Philosophical Axiom, *viz.* *Quæ sunt prima in Compositione, sunt ultima in resolutione: Et quæ sunt ultima in resolutione, sunt prima in Compositione.* "*That which is first in the Composition, is* "*last in the resolution: And those things* "*which are last in the resolution, the same are* "*first in the Composition.* The second Argument is grounded upon another axiom commonly received. That is, *Nutrimur iisdem quibus constamus.* "*We are Nouri-*"*shed by those things of which we are con-*"*stituted, or made.* The third argument shall be to shew, and prove a necessity of all Bodies being formed out of water; because neither the four Elements of of the *Peripateticks*, nor the *Tria Prima*, or three Principles of the Chymists, can possibly concur to the constituting of Bodies, as either the Efficient, or Primary Matter; they being themselves but great disguised Schemes of one and the same Catholick Matter, *Water*, from whence they were made, and into which they are ultimately to be resolved, and uniformly to be reduced, either by Art, or Nature. All which assertions I hope to prove, both by Experiment, and Reason, and shall likewise endeavour to strengthen by good and sufficient Authorities.

Section the Fourth.

---

61 AS to the first Argument, founded on that Axiom, that *All Bodies are made of that Matter into which they are ultimately resolved*, and *è Contra*; This Maxim is agreed upon of all hands, both by the *Aristotelians*, the Old Chymists, and the New ones; and that almost upon the same ground. For the first supposed all Bodies reducible at last into Fire, Air, Water, and Earth; and therefore

 held

held the Quaternary of Elements, [*which, by the way, they could never yet sufficiently prove.*] And the Second believed Salt, Sulphur, and Mercury to be the first Principles of all Bodies. And the last sort, the modern Chymists, hold Spirit, Oyl, Salt, Water, and Earth, to be the true Primary Principles of Bodies, for the same reason; *viz.* because many Concrets are resolvable by fire into the first three, if not into the last five, distinct Substances before named.

But that all Bodies are by Art to be 62
brought back, uniformly, into water; hear what that Learned Man, *Helmont*, saith. *Nostra namque operatio Mechanica mihi patefecit, omne Corpus* [*puta saxum*] *Lapidem, Gemmam, Silicem, Arenam, Marcasitam, argillam, terram, Lapides coctos, vitrum, Calces, Sulphur, &c. Transmutari in Salem actualem, æquiponderantem suo Corpori, unde factus est: Et quod iste sal aliquoties cohobatus, cum sale circulato* Paracelsi, *suam omnino fixitatem amittat, tandem transmutetur in Liquorem, qui etiam tandem in aquam insipidam transit: Et quod ista aqua æquiponderet sali suo, unde manavit. —— Plantam verò, carnes, ossa, Pisces, &c. quicquid similium est,*

Helmont *in Tract. de Elementa. ß. 11, 12. p. 43. & De Terra. p. 45. ß. 15.*

eſt, novi redigere in mera ſua Tria, unde poſt modum aquam inſipidam Confeci; Metallum autem, propter ſui ſeminis anaticam commiſtionem, & arena [quellem] difficilimè in ſalem reducuntur. Cum igitur arena, ſive terra Originalis, tam Arti, quam Naturæ reſiſtat, nec queat ullis [unico duntaxat Gehennæ artificialis igni excepto] Naturæ vel artis, à primæva ſui conſtantia recedere; ſub quo igne artificiali, arena ſal fit, at tandem aqua; quia vim habet agendi ſuper ſublunaria quævis abſq; reactione, &c. "*For our handy-craft Operation [that is, his Liquor,* Alkaheſt*] hath manifeſted to me, that all Bodies [to wit, the Rocky Stones] the Pebble, the Precious ſtone, the Flint, Sand, Marcaſits, Clay, Earth, Brick, Metal, Glaſs, Lime, and Brimſtone, &c. may be reduced into a real Salt, equal in weight to its own Body from whence it proceeded: And that Salt being often cohobated with the circulated Salt of* Paracelſus, *doth altogether loſe its fixedneſs, and is tranſmuted into a Liquor, which alſo at length becomes inſipid water; and that water is of equal weight to the Salt of which it was made.—— But Plants, Fleſh, Bones, Fiſh, &c. and every ſuch thing* [ſaith he] *I know how to reduce into its three firſt Principles,*

"*from whence afterwards I have made an*
"*insipid water: but Metal, by reason of its*
"*strict, and exact commixture with its Seed,*
"*and the Sand [quellem]are most difficult-*
"*ly reduced into Salt: for Sand, or the Ori-*
"*ginal Earth, doth resist as well Art, as*
"*Nature, neither will by any means [the*
"*only artificial fire of* Gehenna *excepted;*
"*that is, the* Alkahest] *be made to recede*
"*from its first-born constancy, &c. [un-*
"*der which artificial fire the Sand is made*
"*Salt; and at last water] because it*
"*hath a power to work upon any sublunary*
"*Body, without its re-acting upon it*
"*again.*

Helmont. *Complex. atque Mistion. Figment. p.* 88. *§.* 27.

He likewise tells us, in his Tract, entituled, *Complexionum atque Mistionum figmentum. Novi enim aquam [quam manifestare non Libet, &c. For I know a Water [which it is not fit to discover, meaning the* Alkahest,] *by whose help all Vegetables are changed into a distillable juice, which leaveth no feces in the bottom of the glass: which distilled juice, if it be mixed with* Alkalies, [*or fixed Salts*] *is reduced totally into insipid and Elementary Water.* 63

And a little further in the same Tract, he tells us: *That he took an Oak-Charcoal, and mixing it with an equal weight of the* 64

Liquor

Liquor Alkaheſt, *he put it in a glaſs, Hermetically Sealed, which being kept in a* Balneo *for three dayes, it was in that time turned into a couple of Diaphanous Liquors, of different colours, which ſwam upon each other; which being diſtilled together [in Sand] by a heat of the ſecond degree, it left the bottom of the glaſs as clear, as if it had never been uſed. The two Liquors of the Coal might be diſtilled with the heat of a Bath, but the diſſolving Liquor, [or* Menſtruum] *in that degree of heat would remain at the bottom of the Glaſs, not impaired in its weight, or Virtue. And that the aforeſaid two Liquors of the Coal, being mixed with a little Chalk, at thrice diſtilling, did aſcend of the ſame weight as before; but having loſt all their diſtinguiſhing qualities, it became undiſcernable from Rain-water.*

65 The Operations of this Liquor [which you have heard] in reducing all Bodies uniformly into water, is, I think, of very great force to evince, what I have here affirmed, *viz.* that all Bodies were Originally Water. But after all this ſtreſs I lay upon theſe Experiments of *Helmont's*, it may be objected by ſome, That they not being poſſeſſors of this Liquor, may be allowed to doubt of

the truth of what he hath deliver'd concerning it. To which I anſwer, firſt, that I think it no cogent Argument, to conclude there is no ſuch thing, becauſe many men are not poſſeſſors of it; and if this ſhould be admitted, all other Arts and things, that are poſſeſſed by any Man [and not known to the common people] would be liable to the ſame exception; and every Cobler, or Ploughman would conclude the impoſſibility of the effects produced by moſt Mathematical Automatons, or Engines, becauſe he either knoweth not, or hath not ſeen the contrivance of the thing, or elſe is not able to conceive the reaſon of its Operation: And if every Man [that knoweth more than the Vulgar] would make it his own caſe, they would, I ſuppoſe, think it an unreaſonable and hard way of judging of things.

Secondly, the Man is ſo conſentaneous 66
to himſelf in his Experiments, that that very thing to me appeareth an Argument of his Truth. And as to his veracity in thoſe things he delivers as matter of fact, [and upon his own knowledge,] I do not find that even his Enemies have detected him of Falſhood; and I am ſure, I have hitherto found him moſt true, in

what-

whatſoever he hath delivered us as his own Experience [though poſſibly many of thoſe things do not at firſt ſight ſeem over-probable.] But leſt I may ſeem over-partial, I will give you a Teſtimony of him [that may be *inſtar omnium*] and that ſhall be from a Man, of whom the World is fully ſatisfied, not only as to his candid Temper, bnt alſo of his ability to judge, both of Men, and things; and the unwillingneſs of his Nature to encourage falſhood: and that is the Inquiſitive, and Honourable Mr. *Boyl*, who ſaith thus both of him, and the *Alkaheſt*.

76 *If our Chymiſts will not reject the ſolemn, and repeated Teſtimony of a Perſon* [ſpeaking of *Helmont*] *who cannot but be acknowledged for one of the greateſt Spagiriſts they can boaſt of, they muſt not deny that there is to be found in Nature another Agent, able to analyze compound Bodies leſs violently, and both more genuinely, and more univerſally than Fire: And for my own part, I have found* Helmont *ſo faithful a Writer, even in divers of his improbable Experiments, that I think it ſomewhat harſh to give him the lye, eſpecially to what he delivers upon his own proper Tryal. And I have heard from very credible Eye-witneſſes ſome things,*

Scept. Chymiſt. Carneades *Dialogue* II.

things, and seen some others my self, which argue so strongly, that a Circulated salt [*or a* menstruum, *such as it may be*] *may by being abstracted from compound Bodies, whether Minerals, Animals, or Vegetables, leave them more unlocked than a wary Naturalist would easily believe; that I dare not confidently measure the power of Nature, and Art, by that of the Menstruums, and other Instruments, that even eminent Chymists themselves are as yet wont to imploy about the Analyzing of Bodies.* Thus far he.

Besides, he that had laboured more than thirty years in the fire, and making Experiments, in all probability might attain this secret: since *Geber*, and many of the *Arabian* Philosophers had it before him; as also *Basil Valentine*, *Raymund Lully*, and *Paracelsus*. Nor can I believe so grave and great a man, would in his Old Age, near his Death, impose falshoods and lyes upon the World. 68

But without the assistance of this Liquor, this Doctrine may be made out; though by more troublesome, and tedious wayes; as we shall now proceed to shew. 69

The same worthy man, *Helmont*, saith, [and 70

[and I have found it true by experience] *Olea & pinguedines, per ignem separata; adjecto pauco sale* Alkali, *saponis Naturam assumunt, atque in aquam Elementalem abeant.* [And again, thus :] *Omne* Alkali, *addita pinguedine, in aqueum Liquorem, qui tandem mera & simplex aqua fit, reducitur [ut videre est in sapone, &c.] quoties per adjuncta fixa, semen pinguedinis deponit;* That is, " *That fats, and " Oyls distilled by fire, a little of an* Alkaly, " *[or fixt salt] being added, do become " soap, and at last, may be turned into " Elemental Water.* —— *All* Alkalies, " *fats being added, are converted into watry " Liquors, which at last is made and reduced " into mere simple water [as it is to be " seen in soap, &c.] as often as by a fixed " adjunct,* [such as Chalk] *it shall be " made to lay aside its seed, and fatness.*

Complex. & Mistion. Elem. figm. p 86 §. 12.

71 And again, *Omne Oleum distillatum, in salem est mutabile, & in aquam per adjuncta.* " *All distilled Oyl is to be changed into Salt, and by adjuncts into water.* Alsso, the best spirit of Wine, which is totally inflamable, if it be joyned with salt of Tartar, will be transmuted into mere water : which salt of Tartar it self, by the help of Oyls [as is above declared] will at last be reduced into water.

All

All Vegetables are reducible by distil- 72
lation into Water, Oyl, and Salt; the water cohobating upon Chalk becomes merely Elemental; the Oyl and Salt may, as is said above, be made to unite into a Saponary Body, which distilled, yield a stinking water, which being oft re-distilled from Chalk [or some such Body] having laid aside its seminal qualities, is indiscriminate from common water: The Salt it self [which is accounted the most permanent principle] yet by the help of fire, well contrived Vessels, and proper adjuncts, it may be reduced into a *Volatil Menstruum*, which being put to act upon Bodies, as a dissolvent, it loseth its saline acrimony, and by repeated operations it is totally converted into insipid water.

All Animals upon the face of the 73
Earth are remigrable into water [of which they were formed] And first, as to Snakes, Vipers, Eels, Froggs, &c. these being perfect Animals, as consisting of Organical parts, as Hearts, Stomacks, Livers, Galls, Eyes, &c. [not to mention Worms, and other insects] some of them accounted hot Creatures, and so full of vivacity and life, that several of them will survive after the taking their hearts

hearts out of their Bodies ſome hours, [not to ſay, dayes ;] I ſay one would little ſuſpect by their out-ſide, theſe Creatures ſhould abound with moyſture as they do. For, if any of them be put to diſtillation, you ſhall perceive them to boyl in their own juice, and to afford an incredible quantity of Phlegmatick Liquor, which being cohobated upon dry Bodies, as is directed in the reduction of Vegetables, returneth to water ; alſo their Oyls, and fatty ſubſtances, being joyned with an *Alkaly*, and made into a ſoap, then diſtilled, they yield a ſtinking water, which cohobated, as the other, doth likewiſe return into water.

74 All other ſorts of living Creatures are, by the help of fire, to be diſſected into Oyls, a fixt, and a volatile Salt [though they yield moſt of the latter] an Empireumatical Spirit, and Phlegm : all which by the above-ſaid helps, and the like repeated Operations, will at laſt be brought into water.

75 Middle Minerals, and Mineral Salts, by Art are reducible into Corroſive Spirits ; which acting upon Bodies, are diſpoil'd of their acrimony ; and, at laſt, return to the ſhape of water.

76 As for Minerals, and Metals ; if they be

be fluxed with *Alkalies*, they are thereby rob'd of their Sulphurs; to which if you add Oyl, it is made soap, and then to be dealt with as is above directed, by the Example of both Vegetables, and Animals: or else the Sulphurs of Minerals, separated from the *Alkalizate* Salt, may be burnt, and the Fume caught by a Glass-Bell, [as is usual in making Oyl of Sulphur *per Campanam*] it will be turned into a corrosive Spirit, which will be reduced into water; as I have shewed above, other corrosive Spirits may be by acting upon Bodies.

Metalline *Mercury*, or Quick-Silver, 77
[that peerless body for homogeneity, and likeness of parts] which exceedeth water in weight at least fourteen times, [the parts of it being so forcibly compressed by the power of its Seed] may yet totally be reduced into water, in purposely contrived Vessels, and a skilfull management of the fire; as *Raymund Lully* doth witness, and Experience with him.

Nay, Nature her self doth in time 78
[by the help of Putrefaction, and ferments residing in the Earth,] reduce into water the bodies of Vegetables, and Animals, whether Fish, or Flesh; also Salts,

Salts, Ashes, Stones burnt to Lime, &c. witness the dunging of Land by these things. Nay, Metals themselves in time, having past their ἀκμὴ or prime, degenerate into middle-Minerals, and Salts; and then return to water. So that you see, all Bodies have water for their first Matter; and are by Art and Nature reducible into it again at last.

79 *Paracelsus* [a Person hardly inferiour to any Man in the knowledge of Metals, and Minerals] giveth us his Opinion of the production of Metals, and Stones, from water, in these words. *Sic ergo Mirabili Consilio Deus constituit, ut prima Materia Naturæ esset aqua, mollis, levis, potabilis; Et tamen fœtus seu fructus ipsius est durus; ut Metalla, Lapides, &c. quibus nihil durius est.* "*So therefore* "*God hath ordered, by a wonderful Counsel,* "*that the first Matter of Nature should be* "*water, soft, gentle, potable; and never-*"*theless the off-spring, or fruit of it, is* "*hard; as Metals, and Stones, &c. than* "*which nothing is harder.* Paracel. *Liber de Miner. Tract.* 1. p. 342.

Plato also is of the same judgment with him; for he tells us. *Aquæ genera duo sunt præcipua, unum humidum* ὑγρὸν, *alterum fusile* χυτὸν: "*There are two sorts* "*of Waters, one moyst, the other fusil, or* Plato *Timæ.* p. *Græc.* 488. *Latin.* p. 718.

E "*to*

"*to be melted.* And presently after, he explaineth what he meaneth by *fusil* waters. *Ex his verò quas aquas fusiles appellavimus, quod ex tenuissimis, levissimisque, fit densissimum, uniforme, splendidum, flavumque, & prætiosissima res est, aurum florescens per petram compactum est:* "*But of* "*these, which we call* fusil *waters,* [*or to* "*be melted*] *Gold flowering through the* "*Rock is compacted; for it is, of a most soft,* "*fine, and tender thing, made most hard,* "*uniform, splendid, and yellow, and is a most precious thing.*

The Seeds of Minerals, and Metals 80
are invisible Beings; [as we have shewed, above, the true Seeds of all other
things are;] but to make themselves
visible Bodies they do thus: Having got-
ten themselves sutable Matrices in the
Earth, and Rocks, [according to the 81
appointment of God, and Nature] they
begin to work upon, and Ferment the
water; which it first Transmutes into a
Mineral-juice, call'd *Bur*, or *Gur*: from
whence by degrees it formeth Metals.
To which purpose I shall give you a testimony, or two. The first we bor-
row from that Book, Entituled, *Arcæ*
Arcani artificiosissimi apertæ, beginning
thus: *Igitur Notandum est*, &c. Which
because

because the passage is long, we will only give you in *English*, thus: *Therefore it is to be Noted, that Nature hath her passages and Veins in the Earth, which doth distill waters, either Salt and Clear, or else turbid. For it is alwayes observable by sight, that in the Pitts, or Groves of Metals, sharp, and salt waters distill down; therefore while these waters do fall downwards, [for all heavy things are carried downwards] there doth ascend from the Centre of the Earth, Sulphurous Vapours, which do meet them. Wherefore if so be, the waters be saltish, pure, and clear, and the Sulphurous Vapours pure also, and both of them do strictly embrace each other in their meeting, then a pure Metal is produced; but in defect of such purity,* [that is, of the Water, and Vapour] *then an impure Metal is generated: in Elaborating of which Nature spendeth near a thousand years before she is able to bring it to perfection; and this happeneth either by reason of the impurity of the Salt, Mercurial waters, or the impure Sulphurous Vapours. When these two do embrace each other, shut up close in Rocky places; then by the Operation of Natural heat there doth arise from them a moyst, thick, fat Vapour, which seateth it self where the Air cannot* Theatrum Chym. vol. 6. p. 305.

come, [for else it would flye away:] of
this Vapour a Mucilaginous, and unctuous
Matter is made, which is white like Butter;
Mathesius *calls it* Gur : *it will spread*
like Butter, which I also can shew in my
hand, above, and out of the Earth. And
the same Author again, thus. *The*
Matter of Metals before it be Coagulated in-
Arca Ar- *to a Metalline form, is like Butter made of*
can.p.318. *the Cream of Milk, which may be clam'd,*
or spread as Butter, which he [he mean-
eth *Mathesius*] *calleth* Gur, *which I also*
[saith the Author] *have found in the*
Metallo- *Mines, where Nature hath produced Lead.*
graph.p.50. And that Industrious Metallurgist, *Web-*
ster, [who hath likewise noted the same
passages out of this Author] assureth
us, that he hath in his possession some
pounds weight of this Metalline Liquor,
called *Gur*.

To which I will also add my own Te- 82
stimony; which is, that about eighteen
years past, having made a Visit to a
Friend, who dwelt upon the Borders of
Derby-shire; and who had at that time
newly discover'd a Lead-Mine in his
Ground: I remember, that being at the
said Mine I saw upon the Work-man's
breaking a stone of Lead-Ore, a bright
shineing Liquor spurt forth; which in a
little

little while did coagulate, and become solid.

83 And that Worthy Man, *Helmont*, confirms what we have related of this Metalline juice, in these words:

Non raro nempe contingit, quod Metallarius, in fodinis, saxa diffringens, dehiscat paries, & rimam det, unde tantillum aquæ, subalbidæ, virescentis, manavit, quod mox concrevit instar saponis liquidi [Bur *voco*] *mutatòque deinceps pallore subviridi, flavescit, vel albescit, vel saturatius viridescit. Sic enim visum est, quod alias intus, absque saxi vulnere, fit; Quia succus ille interno Efficiente perficitur. Est ergo prima seminis Metallici vita in Condo, sive Promptuario loci, homini plane incognita: at ubi semen in lucem, Liquore vestitum, prodit, Et gas incæpit Sulphur aquæ inquinare, vita est seminis media; ultima verò, cum jam indurescit*; that is, "*It ma y times happeneeh, that a Mine-Man, in the Pits, breaking stones, the wall is opened, and a Chink is made; from whence a little water hath flowed, of a whitish greeness, which presently hath thickned like soft soap.* [*I call it* Bur, saith he, but I suppose it should be written *Gur*] *and by and by the somewhat greenish paleness being changed, it* "*groweth*

Helmont *in Magn. Oport.* p. 127.

“ *groweth yellowish, or whitish, or more* “ *fully greenish: So that that is brought* “ *to sight, which nevertheless was made* “ *within, if the stone had not been broken;* “ *because that juice* [*or Liquor*] *is* “ *brought to perfection by an internal Effici-* “ *ent; therefore the first life of the Metal-* “ *lick seed is bid in the little store-house,* “ [*or Cellar*] *of the place, altogether* “ *unknown unto Man: but when the seed* “ *is brought to light, invested with a Li-* “ *quor, and the* Gas *hath begun to defile* “ *the Sulphur of the water, it is then the* “ *middle life of the seed; but the last life* “ *is, when it is now grown hard,* [that is, “ become a true Metal.]

And again, that this Metalline juice, 84
which he calls *Bur,* [and other Authors *Gur,* which is the true proximate Matter of all Metals] was Originally nothing but water, coagulated by the power of Metalline Seeds; Hear what the same Author sayes: *In terra nimirum fracescens aqua, semen locale vel insitum acquirit, ideoque vel in Liquorem,* [*Leffas*] *ad omnem Plantam, vel in succum* [Bur] *Mineralem transit, juxta species, per directionem seminum Electas:* “ *Indeed the* “ *water, by continuing in the Earth, grow-* “ *ing putrid, doth obtain a local, or* “ *im-*

Helmont, *in Element.* p. 43.

" *implanted Seed ; and by that means it is* " *changed either into the Liquor* [Leſſas,] " *for all Plants* [*to be made out of it*] *or* " *elſe into the Mineral juice* [Bur] *ac-* " *cording to the particular kinds, choſen by* " *the direction of the Seeds.*

85 But that you may not think, that Lead alone is formed from this Buttery, or Soap-like ſubſtance, which we have been ſpeaking of ; but alſo that all other Metalline, and Mineral bodies are produced from the ſame ; I ſhall give you an Inſtance, or two ; *Eraſtus*, as I find him quoted by *Webſter*, ſaith, *I have two* Metal. p. 44. *ſtones of Iron, one of them of an Ironiſh colour, the other of the colour of the ſhell of a ripe Cheſnut ; altogether ſoft, and fatty, that may like Butter be wrought with the fingers ; from which, notwithſtanding, hard, and good Iron was extracted by the fire.*

86 Concerning the generating Silver from ſuch a Mineral-Liquor, that Honourable Perſon, Mr. *Boyl*, tells us [from Scept. Chym. p. 360. *Gerrhardus*] thus. *Item aqua Cærulea inventa eſt* Annebergæ, *ubi Argentum adhuc erat in primo Ente, quæ coagulata, reducta in Calcem fixi & boni argenti :* " *Alſo that at* Anneberge *a blew water* " *was found, where Silver was yet in its* " *firſt Being, or Ens, which coagulated,*

"*was reduced into the powder, or Calx of* "*fixed and good Silver.*

As for Gold, and Antimony, *Paracelsus* saith, it is to be found in its *Ens primum*, or first Being, Liquid, and in the form of a Red Liquor, or Water, which afterwards is coagulated and exalted into Gold.

Paracelsus, *in lib. de Ren. & Restor. p.* 43, 44, 45 *& Chyrurg. Mag. p.* 117 243, 244. *De Renov. p.* 45. *Rer. Natur. Lib.* 8. *p.* 104.

Again, he sayes of the *primum Ens lege* 87
solis, that it is a fugacious Spirit, as yet consisting in volatility, as an Infant in the Womb of a Woman, and is sometimes like a Liquor, and sometimes it is found like an *Alcool*, or subtile powder.

'Tis a common known thing, that 88
those Men which bore the Ground to find out Coal-mines, do, when they come near the Mine, bring up in their borer a sort of matter they call Soap-stone, which is like fat Clay, but of a black colour, and will, when new taken out of the Ground, spread like butter, as *Gur* will do; but in the Air will soon become so hard, that it will not cut with a Knife:

I might here take notice of what *Ru-* 89
landus hath said of the *Medulla Lapidis*, which the *Germans* call **Steinmarck**; some of which is white, some red, and some of other colours; and

and most of it in substance like the forementioned *Gur* : but to avoid being tedious, I forbear. And of this sort of coagulated water were those Pebbles made, which *Peireskius* found soft under his feet in the River *Rosne*; as is related in the tenth History of Petrification, in the first Section of this Discourse.

90 So that, I think, it is evident, first, by the afore-cited Authorities, which hold that all bodies were made of water, and seed; and secondly, by the alleadged Experiments, teaching the Reduction of all bodies into water again; that the Original of all Concrets, [even those solid ones of Metals, and Stones] is water.

91 And I do not find that very ingenious man, Mr. *Boyl*, to be against this Opinion: for he saith thus; *Yet thus much I shall tell you at present, that you need not fear my rejecting this Opinion; since however the* Helmontians *may in Complement to their Master, pretend it to be a new discovery, yet though the Arguments be for the most part his, the Opinion it self is very Ancient.* Scept. Chym p. 218.

92 I have now done with the first Argument, that is, that all Bodies are made of

of those things into which they are at last to be resolved, and that I have proved to be water.

I now proceed to the second Argument, *viz.* that all Bodies are Nourished by that of which they are Constituted. 93

Section the Fifth.

THat Vegetables are nourished by water, will plainly appear from hence, that no Plants do either grow, or increase without the assistance of water; either by the way of Rain, or Dew, or else by the overflowing of some Spring, or River; for if they be destitute of water, they dye, and wither. 94

And it is commonly known, that the tops of Rosemary, Marjoram, Mint, Baume, Penny-ryal, Crows-foot, and many other Plants, will thrive, flourish, and grow to a large Bulk [without being Planted in the Earth,] if they be only put into a Glass with fair water in it; into which they will shoot out springy Roots, and from whence they will gather 95

suf-

ſufficient Nouriſhment to become large Plants.

96 To confirm which I ſhall relate a couple of very remarkable paſſages; the one borrowed from that honourable Philoſopher, Mr. *Boyl*; the other from that Learned Naturaliſt, *Helmont*.

97 Mr. *Boyl* tells us, that he cauſed a certain quantity of Earth to be digged up, baked in an Oven, and weighed; and then put into an Earthen Pot, in the which he ſet the ſeed of a Squaſh, which grew very faſt, [though planted too late, *viz.* in the Moneth of *May*] it being watered only with Spring, or Rain-water: in *October* [by reaſon of the approaching Winter] he cauſed it to be taken up, and the weight of it, with its ſtalk, and leaves, was found to be two pounds, twelve Ounces; and the Earth [in which it grew] being baked as before, it was found to be exactly the ſame weight. *Scept. Chymiſt. p.*

98 *Helmont*'s Relation is this: He took, he ſaith, two hundred pounds weight of Earth, which was dryed in an Oven, and putting it into an Earthen Pot, he moyſtened it with Rain-water, and in it he Planted the trunck of a Willow-Tree, which weighed five pounds, [covering the *Complex. & Miſt. fig. p. 88. ſſ. 30.*

the Pot with an Iron cover, which had a hole for the Tree to grow out at,] and at the end of five years, he took up the Tree, and found it to weigh one hundred sixty nine pound, three Ounces; and the Earth being dryed, was of the same weight as at first.

Now if this be throughly consider'd, 99
from what can we possibly suppose, the
bulk of the Swash, and this great addition of 164. pounds weight to the Tree,
did proceed but from meer water; there
being nothing else added to either of
them? and no doubt, Nature observeth
the same course in producing all other
Vegetables; whether springing up from
their innate Seeds, or transplanted into
other soyls: for the Earth is only a Receptacle to receive the seeds of things,
[and to sustain the weight of Minerals,
Animals, and Vegetables: which Seeds
conceive in the water; where they be-
get themselves Bodies, and from which
all Plants arise; and by the power of the
Architectonick Spirit of the seed, fermenting the particles of water, do proceed the stalks, wood, leaves, flowers,
fruit, grain, [or Casket of the real
seed] as also the Colours, Odors, Tastes,
and all the specificate qualities of the
Plant,

Plant, according to the *Idea* wrapt up in the boſom of the ſeed. Animals alſo are nouriſhed by water; ſome immediately, others mediately.

100 Immediately, from meer water, as Salmon, Sturgeon, and ſeveral other ſorts of Fiſh, in whoſe ſtomacks no food, that I know of, was ever yet found. And to confirm this; *Rondeletius* [an Author of good credit] affirms, that his Wife kept a Fiſh in a large glaſs, and fed it with nothing but water [ſo long] till it grew ſo big, that it could no longer be contained in the glaſs; which they were forced to break to get it out.

101 Thoſe living Creatures that are nouriſhed immediately by water and Vegetables, are moſt ſort of Cattel proper for food; ſo that in theſe Beaſts, which feed upon Corn, Graſs, and other Herbs, [which are really but water, once removed from its primitive ſimplicity by the power of Seeds,] water is a ſecond time tranſmuted, by the Ferment of a Beaſts ſtomack, by which it is changed into Chyle, Blood, Milk, Urine, Fleſh, Bones, Fat, Sinews, &c. and all theſe different one from another, according to the ſpecies of the Beaſts that feed upon them.

Now

Now these Creatures, and their parts 72
[as the flesh and milk of beasts] serve for food to those Animals that are nourished mediately from water; such are Men, and divers Wild beasts, who live upon the flesh, milk, and blood of Cattel, and by the Ferments of whose stomacks these things are again Transmuted into another kind of Chyle, blood, flesh, bones, milk, Urine, &c. which juices of our bodies are still but water, disguised by the operation of different seeds, and Ferments; which is quickly discovered by distilling them: for, if our blood be distilled, five or six parts of seven will rise in Phlegm [which is easily reducible into simple water, as we have shewed in the last Section before this.]

Nay, the sperm of Man [by which 102
we propagate our selves;] is nothing but water [Originally] altered by the several Ferments of the body, and circulated in the seminal Vessels.

Upon this Subject there is much good 103
matter to be found in that ingenious man, *Simpson*, in his *Hydrologia*.

It now remains, that we prove the 104
growth, and nourishment of Metals and stones from water: which that we may the

the better do, I think it necessary, in the first place, to discover, whether they do really grow, and increase or no; for some men believe, that God Created them at first, when he formed the world; but that since they do neither grow, nor increase: which error we shall endeavour to confute by several good Observations, taken from approved Authors.

105 Almost all the Mystical Chymists have handled this point so obscurely, that though they have asserted, that metals and stones do grow and increase, and that they are generated from a seminal principle; yet have they proved nothing clearly; but left it as a principle to be granted, without any further dispute.

106 'Tis a known truth in *Cornwall*, that after all the Tin, that could be found in a Mine, hath been taken out, and the Mine filled up with Earth; yet within thirty years they have opened them again, and found more Tin generated: of which Dr. *Jordan* doth take notice also, and in the above-cited place he sayes thus: *The like hath been observed in Iron, as* Gandentius Merula *Reports of* Ilna, *an Island in the* Adriatick *Sea, under the* Venetians, *where*

Nat. Bath. Cap. 11. p. 51, & 52.

where Iron is bred continually, as faſt as they can work it; which is confirmed alſo by Agricola, *and* Baccius. *The like we reade of at* Saga *in* Lygiis, *where they dig over their Mines every ten years.*

And of Ilna *it is remembred by* Virgil; *who ſaith,* Ilnaque inexhauſtis Chalybum generoſa metallis. John Matheſius *giveth us Examples of almoſt all ſorts of Minerals, and Metals, which he had obſerved to grow, and regenerate. The like Examples you may find in* Leonardus Thurniſſerus; Eraſtus *affirms, that he did ſee in St.* Joachim's *Dale, Silver grow upon a Beam of wood, which was placed in the Pit to ſupport the work; and when it was rotten, the Work-men coming to ſet new Timber in the place, found the Silver ſticking to the Old Beam. Alſo he reports, that in* Germany *there hath been unripe, and unconcocted Silver found in Mines, which the beſt workmen affirmed would become Silver in leſs than thirty years. The like* Modeſtinus, Fucchius, *and* Matheſius, *affirm, of unripe, and liquid Silver; which when the Workmen find, they uſe to ſay, we are come too ſoon.*

In Sarept. Conc. 3. p. 11, &c.
Alchym. Mag. De Metallis. p. 17, & 19

Lex. Alchym. p. 56.

And *Rulandus* ſaith [ſpeaking of Sil- 107
ver that is to be found Naturally purified

in

in the Mine;] *Sed hoc argentum purum tenuissimis bracteis amplectitur Lapidem; interdum etiam præ se fert speciem Capillorum, interdum virgularum, interdum globi fert speciem, quasi filis convoluti candidis, aut rubris; interdum præ se fert speciem arboris, Instrumenti, Montium, Herbarum, & aliarum rerum.* "*And* "*this pure Silver doth embrace the Stone* "*with most fine Plates; it sometimes also doth* "*bear the shape of hair, sometimes of little* "*twiggs, sometimes of a Globe, as though* "*wrap'd about with thred, white, or red;* "*sometimes it appeareth in the shape of* "*a Tree, Mountain, Instrument, Herbs,* "*and of other things.*

108 Mr. *Boyl* tells us from *Gerrharaus*, thus. In Valle *Joachimacæ*, &c. [saith he] *In the Vale of* Joachim, *Dr.* Shreter *is a Witness, that Silver, in the manner of Grass, had grown out of the stones of the Mine, as from a Root, the length of a finger; who hath shewed these veins, very pleasant to behold, and admirable, at his own House, and given of them to others.*

Peter Martyr. Decad. 3. Cap. 8. p. 139. Webster, p. 48.

And to shew you, that Metals do grow even like Vegetables, it is very remarkable what is quoted by *Webster*, out of *Peter Martyr*, Councellour to the Emperour *Charles* the Fifth, in these words: *They have found by Experience, that the Vein of Gold is a Living Tree, and that the same by all wayes spreadeth, and springeth from the Root, by the soft pores and passages of the Earth, putteth forth branches even to the uppermost part of the Earth; and ceaseth not till it discover it self to the open Air; at which time it sheweth forth certain beautiful colours in the stead of flowers: round stones of Golden Earth, instead of fruit, and thin Plates instead of leaves: These are they which are dispersed through the whole Island* [he is speaking of *Hispaniola*] *by the course of the Rivers, Eruptions of the Springs out of the Mountains, and other falls of the Floods: for they think, such grains are not ingendered where they are gathered, especially on the dry Land, but otherwise in the Rivers. They say, that the root of the Golden Tree extendeth to the Centre of the Earth, and there taketh nourishment of increase; for the deeper that they digg, they find the trunck*

trunck the bigger, as far as they may follow it for abundance of water, springing in the Mountains: of the branches of this Tree, they find some as small as a thread, and others as big as a mans finger, according to the largeness, or streightness of the Rifts, and Clefts; they have sometimes lighted upon whole Caves, sustained, and born up, as it were, by Golden Pillars, and this in the way by which the branches ascend: the which being filled with the substance of the Trunck, creeping from beneath the branch, maketh it self way, by which it may pass out. It is oftentimes divided by incountring with some kind of hard stone; yet is it in other Clefts nourished by the exhalations and Virtue of the Root.

110 To which I might add what *Fallopius* saith of Sulphur, [*viz.*] *Sunt enim loca, è quibus si hoc Anno Sulphur effossum fuerit, intermissa fossione per quadriennium, redeunt fossores, & omnia Sulphure, ut antea, rursus inveniunt plena:* "*For there are places, from whence if this* "*Year the Sulphur be digged out, and for-* "*bearing to dig, by the space of four years,*

"*the Mine-men return, and find them* "*all full of Sulphur, as before.*

And that Salt-Petre groweth, and 111
increaseth, our common Salt-Petre-men will justifie; for after they have extracted all the Salt that they can get out of the Earth that yieldeth it, in two or three years after, they work the same Earth [which for that purpose they carefully lay up] over again; and it yields them a considerable quantity of Salt-Petre, as before.

And concerning Table-Salt, *Matthias* 112
Untzerus produceth many Testimonies from credible Authors, that besides that which is made of Salt-Springs, there are in *Spain*, the *Indies*, and divers other parts of the World, large Mountains of Salt, which as fast as they can be digg'd, grow again, and are quickly filled with Salt.

Untz. de Sale. Cap. 7. p. 33, 34, & 35.

And for Lead, [besides what *Galen* 113
observeth of its increase, both in bulk, and weight, by being kept in a damp Cellar,] *Boccatius Certaldus*, as he is cited

cited by Mr. *Boyl*, saith thus of its growth: *Fesularum Mons*, &c. Of the Mountain of *Fesula*, a Village near *Florence*, that it hath Lead-stones; which if they be digg'd up, yet in a short space of time they will be supplied afresh, and generated anew. I might instance in many more particulars, but I think these sufficient.

114 That Stones do grow, and are made since the Creation, every mans Observations will sufficiently acquaint him: And the Histories cited in the first Section of this Discourse do confirm; and that they are nourished by water, is apparent from the Scituation of Rocks in the Sea, the production of Pebbles in the bottom of Rivers, and that both Mountains, and also gravelly places, are never destitute, or unaccompanied of Springs and Rivulets.

115 And *Paracelsus*, I remember [somewhere] giveth us this Experiment, to prove that stones do grow, and are nourished by water; *viz.* that if a Flint, or Pebble be put in a glass Vessel, and Rain,

or Spring-water put upon it, and distilled from it, if this be often repeated, it will cause the stone to grow so bigg, that at last it will fill up the Glass that contained it.

That Metals, and Minerals are nourished by water, is more than probable 116
from hence, that no considerable Mines are found without a great conflux of waters; which the Work-men are forced to make drains and Pumps to carry away, that they may work dry.

And there is an Experiment, written 117
by Monsieur *De Rochas* [a considerable *French* Author, and Transcribed from him by the Honourable, Mr. *Boyl*] which I shall here insert. *Having* [saith he] *discerned such great wonders by the Natural Operation of water, I would know what might be done with it by Art, imitating Nature; wherefore I took water which I well knew not to be compounded with any other thing than the Spirit of Life; and with a heat artificial, continual, and proportionate, I prepared it, and disposed it, by graduations*

ons of Coagulation, Congelation, and Fixation, untill it was turned into Earth; which Earth produced Animals, Vegetables, and Minerals: The Animals did eat, move of themſelves, &c. and by the true Anatomy I made of them, I found they were compoſed of much Sulphur, little Mercury, and leſs Salt: the Minerals began to grow, and increaſe, by converting into their own Nature one part of the Earth; they were ſolid, and heavy; and by this truly demonſtrative Science, namely, Chymiſtry, I found they were compoſed of much Salt, little Sulphur, and leſs Mercury.

According to this Experiment, Minerals were Generated out of, and nouriſhed by water.

118 From what hath been related, both in this and the fore-going Section, concerning the growth, increaſe, and Vegetability both of Metals, Minerals, and Stones; as alſo concerning thoſe Mineral, Metalline, and ſtony juices, called *Gur*, [or *Bur*] Soap-coal, and the *Medulla Lapidis*, &c. I think it will

 appear,

appear, that both Metals, and Stones, are made, do grow, and are nourished, daily, and at this time; and that from water, of which they were at first made, by the power of their Seeds: And this is the reason, that Metals, and Mines are now usually found in those places where for many Years before there were none; as both *Sandivogius*, and *Helmont* assure us. *Inde fit, quod hodie reperiantur Mineræ in locis ubi ante mille annos nullæ fuerunt*: "*From* "*hence it is come to pass, that Minerals* "*may be found in places, where before a* "*thousand years since, there have been* "*none*. And *Helmont*, thus: *Loca enim quæ fodinis Caruêre olim, suo quandoque die, Maturato semine, fœnora reddent, ditioribus non imparia; quia radices, sive fermenta Mineralium, sedent in loco immediatè, ac in dierum plenitudinem sine fastidio anhelant: quam ubi semen complevit, tum Gas obsidens aquam ibidem, semen à loco suscipit, quod aquæ sulphur dein impregnat, aquam condensat, atque sensim aquam Mineralem transplantat*: "*For* "*places which have wanted* [*or had no*] "*Mines in times past, will in their own* "*time,*

Nov. Lum. Chym. Tract 4. *p.* 314.

Helmont, *In Mag. Oport. p.* 127. *ss.* 39.

"*time, their Seed being ripened, restore*
"*Usury, equal to the richer sort [of*
"*Mines] because the Roots, or Mineral*
"*Ferments, are seated immediately in the*
"*place; and their full time being come,*
"*they [pant] or breathe without [weari-*
"*ness] or loathing: and when it hath*
"*gained a compleat Seed, then the Gas*
"*which is seated in the water of that place,*
"*receiveth that seed of the place, which af-*
"*terwards begets the Sulphur of the water*
"*with Child; condenseth the water, and*
"*by degrees turneth, or transplants it into*
"*a Mineral water.*

119 And, to conclude this Section, I will give you the Judgment of that great Naturalist, *Helmont*, by way of confirmation; because I find him exactly to correspond with all that I have hitherto delivered.

120 His words are these, which you shall find in his *Imago Fermenti*; which because they are long, I will only give you their sence in *English*. *And indeed because the Schools have been unacquainted with Ferments, they have also been ignorant,* *that*

Helmont, *Imag. Ferment.p.94. §§.29,30, 31.*

that solid Bodies are framed only of water, and Ferment: for I have taught, that Vegetables, and Grain, and whatsoever Bodies are nourished by them, do proceed only from water: for the Fisher-man never found any food in the stomack of a Salmon; if therefore the Salmon be made of water only, [even that of Rivers] he is also nourished by it. So the Sturgeon wants a mouth, and appeareth only with a little hole below in his Throat, whereby the whole fish draweth nothing besides water. Therefore every Fish is nourished, and made of water, if not immediately, yet at least by Seeds, and Ferments, if the water be impregnat therewith. From the Salt Sea every fresh Fish is drawn; therefore the Ferment [of the Fish] turneth Salt into no Salt, or at least water into it self. Lastly, Shell-fish do form to themselves stony shells of water, in stead of Bones; even as also all kind of Snails do; and Sea-Salt, which scarce yieldeth to the force of a very strong fire, groweth sweet by the Ferment in Fishes; and their flesh becometh volatile: for, at the time of distributing the nourishment, it is wholly dissi-

dissipated, without a residence, or dreg. So also Salt passeth over into its Original Element of water; and the Sea, though it receive salt Streams, yet is not every day increased in saltness. So the most unmixed, and most purest water, under the Equinoctial Line becometh hory, and stinketh: strait-way it getteth the colour of a half burnt brick, then it is greenish, then red, and quaketh very remarkably, which afterwards of its own accord returns to it self again: truly this cometh to pass by reason of the conceived Ferment of that place, which being consumed, all these appearances cease. So the most pure Fountain-water groweth filthy, through the musty Ferment of the Vessel; it conceiveth Worms, breedeth Gnats, and is covered with a skin. Fenns putrifie from the bottom, and hence arise Frogs, Shell-fish, Snails, Horse-leaches, Herbs, &c. also swimming Herbs do cover the water, being contented with drinking only of this putrid water. And even as stones are from Fountains wherein a stony Seed exists; So the Earth stinking with Metallick Ferments, doth make out of water,

ter, a Metalline, or Mineral Bur; *but the water being in other places shut up in the Earth, if it be nigh the Air, and stirred up with a little heat, it putrifieth by continuance, and is no longer water, but the juice* Leffas *of Plants; by the force of which hory Ferment, a Power is conferred on the Earth of budding forth Herbs. For that putrifying juice by the prick of a little heat doth ascend in smoak, becomes spungy, and is compassed with a skin, because the ferments therein hid require it. Therefore that putrefaction hath the office of a Ferment, and the Virtue of a Seed, and by degrees it obtaineth some measure of Life, and hasteneth by the Virtue of its Seeds into the Nature of Archeuss. Therefore this putrid juice of the Earth, is* Leffas: *from whence springs every Plant not having visible seed, which nevertheless bring forth seeds, according to their destinations. Therefore there are as many rank, putrid, musty smells, as there are proper savours of things. For Odors are not only the Messengers of Savours, but also their promiscuous Parents. The smoak* Leffas *being now comprest together,*

ther, doth first grow pale, then somewhat yellowish, and presently after is of a whitish green colour, and at last fully green. And the power of the several species being unfolded, it gains divers marks, and different colours: in which course it imitates the Example of the water under the Equinoctial Line. Yet in this it differs, that those waters have borrowed too Spiritual and volatile a Ferment from the Stars, and place, without a Corporal bory putrefaction; and therefore through their too frail Seed they presently return into themselves. But Leffas *is constrained to finish the Act, [and obey the Power] of the Conceived Seed. Therefore Rain Conceiving a bory Ferment, is made* Leffas, *and is sucked in by the lustfull Roots:* 'Tis *experienced also, that within this Kitchin [of the Root] there is a new bory putrefaction produced by the Ferment which is* T*enant there; by and by it is brought from thence to the Bark [which is as it were the Liver of the Plant,] where it is inriched with a new Ferment of that part, and is made a Herby, or Woody juice; and at length it being come to*

Matu-

Maturity, it is made Wood, an Herb, or becometh Fruit. If the Arm, or Stem of a Tree *shall be putrefied under the Earth, then the Bark or Rinde becometh dry, and cleaveth assunder, and sendeth forth a smoak by its own Ferment, which in the beginning is spungy, but at length hardens into a true Root: and so Planted Branches become* Trees *by the abridgment of Art.*

Therefore it is now evident, there is no mixture of Elements, and that all Bodies primitively, and materially are made of water, by the help of Seeds, and their Ferments; and that the Seeds being worn out, and exhausted by Acting, all Bodies do at length return into their Ancient principle of water yea, that Ferments do sometimes work more strongly than fire, because that fire can turn great stones into Lime, and burn Wood into ashes, but there it stops; but notwithstanding, if they shall assume a Ferment in the Earth, they return into the juice of Leffas, *and at last into simple water. For Stones, and Bricks,* 121

do

do of their own accord decline into Salt-petre. Lastly, Glass which is unconquered by the fire, and uncorrupted by the Air, in a few years putrifieth by continuance [in the Earth] and undergoes the Laws of Nature, &c.

122 Having now gone through the two first Arguments, by which I proposed to prove the Doctrine I have asserted, which Arguments were grounded on two generally received and allowed Axioms, [*viz.*] Those things which are the last in the resolving, [or retexing] of a Body, the same are found to be the first in its composition. Secondly, we are nourished by those things of which we are made, [or consist.] And having, I hope, sufficiently proved by both of them, that Water is the Original Matter, and Seeds the Efficients of all Bodies; I am now come to the third, and last Argument, which was to shew, and prove a necessity of all Bodies being formed out of water; because neither the four Elements of the *Aristotelians*, nor the three Principles of the Old Chymists, no, nor yet the five of the Modern

dern Chymists, can possibly concur to the constituting of Bodies, as either their Primary Matter, or Efficient; they being themselves but great disguised Schemes of one and the same Catholick Matter, Water, from whence they themselves were made; and into which they are ultimately to be resolved, and uniformly to be reduced.

Section

Section the Sixth.

AND First for the *Chimical* Principles; I have shewed [in the Fourth *Section* of this Discourse,] That the Oyls of *Vagetables*, and their Fermented Spirits, which are their *Sulphurs*; that the Fats, and Oyles of Animals, which are their *Sulphurs*, and also the *Sulphurs* of Minerals, and Mettals, are all of them reducible into Water: As are also both Mineral, Animal, and Vegetable Salts. And as to the *Mercury* of Animals, and Vegetables (improperly enough so called,) they being but of a loose Contexture, are easily made to *remigrate* into water; (as I have taught in the same place:) As also is [though with somewhat more reluctancy, because of its strong Compression by its Seed,) true Mettallin *Mercury*, or *Quicksilver*, as my own experience hath assured me: Which is also confirmed by *Raymundus Lullyus*, the ingenious Mr. *Boyl*, and divers others.

126 All this may be performed two ways, that is, Either by the means prescribed in the forecited pages, or else more solemnly, speedily, and universally, by the help of that rare Solvent, the *Alkahest*: The manner of whose operating upon Bodies, I have described from the relation of that worthy man *Helmont* [in the fourth Section.]

127 Now as to the two other Principles added by the Modern *Chymists*; the one of them, *viz.* Earth, doth properly belong to the School of the *Stagyrit*; and therefore I speak to that, when I come to discourse of the four supposed Elements of Bodies.

128 But as to the other, *viz.* Spirits; they are all of them of one of these two Classes; either Vinous, and made by Fermentation; or Saline, and made without.

129 Now for the Vinous, they are totally inflamable Bodies; and therefore to be Ranked under the Classis of *Sulphurs*; and may be reduced to water, as I have shewed you above: Other *Sulphurs*, and Spirit of Wine it self may.

130 The other sort of Spirits, *viz.* Saline, are nothing but Volatile Salts, diluted with

with Phlegme or water; and therefore by repeated distillations, and careful rectifications, will be brought to constitute a *Lump* or *Mass* of dry Salt: Wherefore it is not an other Principle, distinct from the former *three* of the Old Chymists; and by the same handycraft-means may at last be reduced to water, as I have before shewed the three Principles of the Chymists may be.

131 Nor indeed can any of these three Bodies, called Salt, Sulphur, and Mercury, pretend to be the principles of all Concretes, except only Mercury, or Water; for it is proper for Principles, that they be Primary, and not further resolveable into more simple parts: But both Salts, and Sulphurs [as I have made out above] Being further reducible, *viz.* into Water; they therefore cannot [whilst such] deserve the Name of Principles.

132 Besides, it is very much questioned by those two great Phylosophers, *Helmont*, and *Boyl*, whether the Fire indeed be an *adequate* and fit instrument to Anatomise Bodies? And whether or no those distinct Schemes, into which the common Chymists resolve the matter of Bodies by Fire, [and which they call their three

Principles] were indeed really *existing*, in those Bodies, from which they were Educed; [that they were matterially there, no man will deny; they being themselves composed of water?] But whether they were resident in the Concrete that yielded them, in the same Figures, and Shapes, that the Fire Exhibites them to our Sences, is very disputable? And it may easily be imagined, that the Fire acting upon a Body that it can master, [for some it cannot] doth not only put the small parts, of which that Body consisted, and which were before [in some measure] at rest amongst themselves, into a tumultuous motion; by means of which, they are sent hastily off into the Receiver; but doth also break by forcing them asunder, those small particles of that body into other Shades, Figures, and Sizes: upon which account they do convene together after new manners; and so the Fire may present us with new Bodies, which were not præ-existent in the Concrete, when first exposed to its Action.

But because this point is throughly, and Learnedly handled both by *Helmont*, and my excellent Friend Mr. *Boyl*, in his Scep-

Sceptical Chymiſt, I ſhall ſpare my ſelf the pains of expatiating upon it; and refer the Inquiſitive to thoſe two Authors, for full ſatisfaction in this point.

133 Only I think it very neceſſary in this place, to examine the Arguments which are brought by a very learned man, and Eminent Phyſitian, to evince the real Exiſtence of the Chymical Principles in Bodies, and to prove that they are not products of the Fire. And I the rather take notice of it here; Firſt, becauſe they are not bare ratiocinations of this Learned mans, but experiments; upon which he hath built very much: And Secondly, ſhould I omit to examine theſe Experiments, [which indeed do ſeem weighty] they might perhaps be produced againſt the Doctrine I defend: And ſome might likewiſe object, that I had not dealt candidly with the Chymiſt; in that I had taken no cogniſance of the beſt weapon they have to defend their Cauſe.

134 This Learned man then intending to prove the real exiſtence of Salin and Sulphurous Principles in Bodies, before the action of the Fire upon them, produces Experiments neverthelefs, that are made

Dr. *Willis de ferm.cap* 20. p.16.

by the Fire. His ſence is this: For the firſt, [viz. *Salt*] *it is commonly known, that if the Salt be once waſhed out of the Aſhes of any vegetable, if they be again calcined, they will yeild no more Salt. Moreover, if any concrete being diſtilled, ſhall yeild a very ſharp, and acid Liquor, their Calces* [*or Aſhes*] *do remain leſs Salt*; *and* è contra, *that is, where the Salt is volatized, and become a Liquor, and doth aſcend by the Alimbec, you ſhall in vain ſeek for it in the* caput mortuum: *That which vindicates the Exiſtence of the Principle of Sulphurs in Vegetables, is this*; *Take* Guajacum, *or any other ſort of heavy wood, in pieces or ſhavings, and putting it into a Glaſs-Retort, diſtill it by degrees*; *and it will give you, together with a ſower Liquor* [*which is the Saline* Latex] *a blackiſh Oyl* [*which is its ſulphury part*] *in a great quantity. That this was at firſt in the diſtilled Body, and not all produced by this* ἐγχείρησις, *appeareth from hence, because if you do proceed another way, by which the Sulphur may be taken from the concrete, before it be diſtilled, the Liquor which cometh forth, will be almoſt totally deprived of its Oylyneſs: Wherefore, if you ſhall pour ſpirit of Wine upon the Sha-*

vings

vings of this wood, this menstruum *will extract a great quantity of pure Rozin from it, which is the same Sulphury parts; and if afterwards you take these Shavings that are left, and wash them with common water, and being dry, put them in a Retort, and distil them [as at first] you shall have but a little Oyl. But that which is more to be wondred at, and which doth more fully confirm this truth, is, that several Bodies which have little of Spirit, or Sulphur in them [they being for the most part found amongst Volatils] and which chiefly consist of Salt, Earth, and Water, and are separated into these Elements by distillation, which being again mixed together, doth restore us the same sort of mixts, marked with the same sort of qualities as before;* V. G. *if you distil Vitriol in a reverberating Furnace, you shall have a Phlegme, almost* insipid, *which is its watry part: Then a very sower Liquor, or rather a fluid Salt, and in the bottom remains a Red Earth of a pleasant purple Colour: These being rightly performed, if the two distilled Liquors be poured back upon the* Caput Mortuum, *we shall have the same Vitriol as before, revived of the same colour, taste, and almost of the same weight. The like may be done with Nitre, Sea-ſalt, Salt of Tartler, and perhaps, with Alome, and other Mineral bodies, which you may proceed withal with the ſame ſucceſs; ſo that thoſe concrets that conſiſt of fixed and ſtable Elements, may, like Mechanical Engins be taken to pieces, and put together again, without any prejudice.* Thus far he.

135 Firſt then, he ſaith, that if Salt be waſhed from the Aſhes of a vegetable, though the Aſhes be afterwards never ſo much calcined, yet will they yeild no more Salt; and alſo that thoſe things that yeild a ſower Liquor, have little or no fixt Salt in their Aſhes.

136 The matter of fact I do not deny, but the inference from thence, I ſuppoſe I may. For it is no neceſſary conſequence, that a thing was really exiſting in that form, in the body that yeilded it, in the which Art preſents us with it, when ſeparated from the ſaid body: As for Example, who ever believed, that a Cole was ever really Exiſtent, [as a Cole] in wood, any otherwiſe than materially; and it is ſufficiently known, that the Cole is a product of the Fire, which hath diſſipated ſome parts, of which the wood conſiſted and new modified the

rest: From which action of the Fire, the new body of the Cole resulted: From which Cole, if it be fluxed with an Alkalizat-Salt, may be obtained a perfect, true, and totally inflamable Sulphur, no way distinguishable from common Brimstone,[as I have often proved:] Which Brimstone is a body very different from that of Salt,which the same Cole, if burnt to Ashes, will yeild us in the room of this Brimstone. And if it shall be objected, that this Brimstone is the Oyl of the Wood or Plant, which this Learned man is pleased to call the Sulphury Principle, and which he afterwards tells us may be obtained [together with an acid Saline Liquor, upon which it swimmeth] by distillation from *Guajacum*; if this be objected, I desire it may be considered, First, that the Oyl of the wood was before sent off into the Receiver; and that a much greater Stress of Fire is required to burn the wood into a Cole, then is needful to separate all its Oyl from it. And Secondly, that after it hath afforded all the Oyl which the Fire can make of it, yet then at last this Brimstone may be made out of it. And thirdly, that it be taken notice of, that it is not a sufficient ground

ground [nay, that it is a liberty not to be allowed] to give different bodies the ſame denomination, becauſe they agree in ſome one quality: as this Oyl, and the Sulphur do in that of Inflammability, when they differ in ſo many others, as is obvious to every man.

137 And as to that part of the Experiment alledged by this Learned man; in the firſt place, *viz.* that theſe Concrets, which yeild in diſtilling a ſower Spirit, which is [ſaith he] their Salt volatiſed, and brought into the form of a Liquor; and therefore, as he ſaith, in vain to be ſought for in their Aſhes, in which very little will be found: It proveth no more but this, that according as Bodies are differently made up, ſo the Fire acts diverſly upon their Matter: As is to be ſeen in Wax and Clay, the former of which the fire melts, and the laſt it hardens. Nor doth it appear, that this Saline Liquor was ſuch, whilſt it recided in the Concrete, and before the action of the Fire upon it; any more than it doth, that there is really, and actually reſiding in the body of Wheat, or Barly, before they be made into Mault, and afterwards Brewed and Fermented, a vinous, and in-

inebriating Spirit: Which when they are so managed we find there is. But if otherwise these grains of Barly, or Wheat, shall be ground into Flower, and made into Bread, they then become wholesome Food; of which a great quantity may be eate without procuring drunkenness, which their fermented liquors will cause. And yet from this very substance of the Grain, which affordeth two such bodies, as Drink, and Bread; by a different managing of it, may be made a liquor which is so far a Corrosive, that it will draw Tinctures, [which are solutions of the small parts of bodies] from divers Minerals, Mettals, and Stones, and that many times without the help of External heat. Nor can it with more Justice be affirmed, that these Salts [whether fixt, or volatile] were really and in that form, existing in the wood, or other Concrete; then it may be said, and believed, that there is actually in Bread-corn, the Flesh, Blood, Bones, Sinews, Hair, Nailes, *&c.* of a man; because we see, that by the action of a humane stomach, these things are made out of Bread.

And as to what is alledged concerning the

the Oyl of *Guajacum*, it yieldeth if it be distilled *per se*, but if it be infused in Spirit of Wine, it will impregnate it with a certain *Rozin*, or *Gum*. And the wood after this Extraction, if it be committed to distillation, will not then afford the same quantity of Oyl as before it would have done: That I easily grant, but then it will quite destroy the inference for which this Learned man brings it; *viz.* That Oyl was in that form a constituant Principle of the mixt. For there is a vast difference betwixt *Rozin*, and Oyl; the one being a firm body that will admit of pulverisation, the other a fluid, and unctious body. And besides many other specifical differences, [which, not to be tedious, I purposely omit,] The *Rozin* is a product of Nature, the Oyl, of the Fire. For the *Rozin* or *Gum*, is to be seen in the wood before distillation; and is only taken up, and dissolved in the Spirit of Wine, which being evaporated, it appears again in its own form. But the Oyl is, I grant, substantially, and materially the same with the *Rozin*; and therefore, that being for the greatest part, or totally taken away, the Fire produceth either lesse, or no Oyl: Because if the

Rozin be left in the wood, when it is committed to the Fire, the Fire doth ſpread abroad, break, and new alter the texture of the *Rozin*, and elevating, and making a new combination of its parts; it conſtitutes that Body which we call Oyl; which is in this caſe a real and new product of the Fire, and was not before formally Exiſting in that Body.

138 And it is plain, beſides the inſtances before cited, that by a different mannagement of one and the ſame Concrete, I will cauſe the Fire to Exhibite very different ſubſtances from it; as for Example, take any herb, as Wormwood, Mint, &c. and having bruiſed them, add Yeſt to them, or by any other means, procure a fermentation in the Matter; and then commit it to diſtillation, it will afford you an Oyl, and a vinous Spirit [which rectified, are both of them totally inflamable] but if the ſame herb be bruiſed, and ſuffered to lie upon the Flore ſome dayes, without fermenting, and if it be thus put to diſtillation, inſtead of yielding a vinous Spirit, and an Oyl, as the other did; it will afford an urinous or Armoniack Spirit; which being carefully rectified, will coagulate totally into a maſs of Salt; and

that

that every man knows, is very different both from an Oyl, and a vinous Spirit: For this Salt is not only brittle, but alſo abſolutely uninflamable.

139 And Laſtly, as to what this Author inſtances, concerning Vitriol, Saltpeter, Tarter, and Alome, yeilding of Saline Spirits, which being poured back upon their *Caput Mortuums*, do redentigrate; and return to the ſame bodies as they were before. The matter of Fact I allow to be true; but withal, muſt be allowed to ſay, that it proveth not what he brings it for; nor doth evince, that Salt, and Sulphur, are principles in all bodies; for 'tis the effect of their ſeeds, that forms theſe bodies out of water: For Salts ſomtimes are the products of ſeeds; as I have proved from the regular figures, into which theſe Concrete juices do conſtantly ſhoot; as in Section the Second of this Diſcourſe. So that it is not ſtrange, that the ſmaller parts of theſe Saline juices, being by Fire divorced from the groſſer, upon their being put together, do haſtily run into, and lodge themſelves in the cavities of their own bodies, from whence they were forced by the Fire. And to conclude, there are many bodies which

which the Fire cannot force to confess they are constituted so much as of two of the five modern Chymical Principles; as to instance in Gold, Talk, Silver, &c. and yet by the operation of the Alkahest, even these are at last reducible to water, of which they were made by the power of seed; and the afore-said Oyls, Salts, and Concrete juices, are to be all of them returned to water by the means prescribed in the Fourth Section of this Discourse.

140 And here I must again take notice of two things, First, that this Learned Doctors Experiments are all made by the Fire; which of it self alone I deny to be a proper Agent, to Analize bodies, and to discover to us the truth of those principles of which they are constituted; and that for these reasons, because it doth not work uniformly upon all bodies exposed to its action; for, as I have said before, it cannot of it self separate any one of these supposed Principles, from Gold, Talk, Sand, Silver, and many other Concrets; and yet of some other bodies it will frame, not only Oyles, Salt, Spirit, Ashes, [or Earth, as he is pleased to call it] but also a Cole, Brimstone,

and

and at laſt Glaſs: which three laſt, no man I ſuppoſe will imagin, were really exiſting, in thoſe bodies of which they are made; and yet are they made by the ſame Agent, and from the ſame Subject, of which the Fire produced Salts, Oyls, Aſhes, &c. and therefore upon the ſame ground, may as juſtly plead for the prerogative of being the conſtituent principles of bodies.

141 The Second thing I would have conſidered is this: That thoſe different Shapes and Appearances, into which the Fire hath put the matter of any Concrete, *viz.* Salts, Oyl, Aſhes, Spirits, all of them are yet ſo compound, that they may be yet further returned and divided into more ſimple parts; *viz.* into water, which is indeed the only, and true material Principle [deſervedly ſo called,] for it is a primary, and ſimple body, into which at laſt, all Concrets, [and even the other Four ſuppoſed principles of this Learned mans] are reduced both by Art, and Nature; and of which they were made. So that we may truly affirm with the Antient Philoſophers, ἓν εἶναι τὰ πολλὰ, καὶ τὰ πολλὰ ἕν; One is many, and many One.

Plato, Hippo, & Anaxag.

So

So that though this Learned Doctor, ſhewed much witt in building ſo fair and ſpecious a Philoſophical Structure, from theſe five ſuppoſed principles, yet can it be no ſafe dwelling in it; becauſe the Foundation is unſound.

143 I have been the fuller in diſcuſſing the Experiments brought by this great man, in favour of his five Chymical Principles; Firſt, becauſe indeed they have a very fair appearance, till they be throughly examined.

And Secondly, I would be very loath to have it thought, I would endeavour inconſiderately or upon ſlight grounds, to diminiſh the fame this ingenious man hath already gained in the World by his Writings.

And now having examined not only the *Tria Prima* [or three firſt Principles of the Old Chymiſts] but alſo the five Principles of our Modern Chymical Philoſophers; and not being able to allow them the Title of Principles, for the reaſons above alleadged: I will likewiſe examine the Quaternary, or four Elements of the *Ariſtotelians*, and ſee, whether they can plead any better Title to be allowed, and eſtabliſhed, the Principles, or Elements, of which all Bodies are made.

Section the Seventh.

145 THe *Quadriga*, or four Elements of the Peripateticks, hath for a long time gained the priviledge, of being esteemed the constituent Principles of all Concretes: [which therefore are usually stilled compound-Bodies] for they say of Fire, Air, Water, and Earth, all sublunary Bodies are made, and from the divers mixtures of these, do arise all generations, corruptions, alterations, and changes, that happen to all sorts of Bodies.

146 And first for the Element of Fire, [placed by *Aristotle* under the Globe of the *Moon*, but never yet seen by any man,] certainly it is nothing else but Heat; and that we know is caused by the violent and nimble agitation of the very minute-parts of Matter: And though there be Heat, [and consequently a kind of Fire] in the Bodies of Animals, yet this is no radical Principle

ciple; but a product of vital Fermentation. The like of which we see is produced by the fermentation of Wines in the Barrel, to whose Bung, if the flame of a Candle be held, the subtil vapours of the Wine take flame and burn; which vapours, if they be otherwayes debarred of all vent, they by their brisk motion, cause an intense heat; and sometimes burst the Vessels that contain them. And this hapneth not only to Wines, but even to water it self; for it hath been observed in long Voyages [which somewhere is also taken notice of by Mr. *Boyl*] that our *Thames* water, being kept close stopt, assisted by the motion of the Ship, and its own secret fermentation, a Candle being brought near the vent, upon the opening of it, hath set all the Cavity of the Vessel into a flame. There is the like reason for the bursting forth of flame from wett and closely compressed Hay; as also from the Action of dissolvents upon Mettallin Bodies, &c. in which action, if the Glasses be stopt, they break with great violence: From the incoercible nature of which, we may conclude, that Fire [if there were ſuch an Element] can never enter, as a conſtituant Principle, into the Compoſition of Bodies; but it is rather, as *Helmont* ſtiles it, *deſtructor ſeminum*, the deſtroyer of Seeds, and is a fitter Inſtrument to Analize, and take Bodies in pieces, by not ſuffering their parts to be at reſt amongſt themſelves, [to which purpoſe it is generally employed] than to conſtitute any. And therefore in this particular, *Paracelſus* was groſly miſtaken, where he undertakes to teach us a way to ſeparate the Element of Fire from Bodies, and afterwards pretends to make a new ſeparation of Elements from them again. For, if we will ſuppoſe an Element of Fire, yet if that be further reducible, it muſt of neceſſity loſe both the name and nature of an Element.

147 But Fire is but an Accident, [no diſtinct ſubſtance, or radical Principle of Bodies;] for Fire, or Heat, as I have ſaid before, doth reſult from the motion, which the ſmall parts of Matter are put into by the power of their Seeds, and Ferments. For Fire cannot ſubſiſt of it ſelf [as matter can, and doth] but neceſſarily requireth ſome

ſome other Body, to which it may adhere, and upon which it may Act: Which Bodies are either of a Vinous nature, as the fermented Spirits of Vegetables; or their Rozinous, and Brimſtony parts; or elſe of an unctuous, and fatty nature, as the Greaſe, and Fatts of Animals; or elſe of a Bituminous ſubſtance, as the Sulphurs of Minerals and Mettals are. And that all this is but diſguiſed Water, which hath got new textures by the operation of Seeds, and Ferments, I hope I have ſufficiently evinced before. So that without we will much injure Truth, we muſt degrade Fire from being an Element or Principle, in the conſtituting of Bodies.

148 Nor doth *Air* enter Bodies, as an Element of which they are compoſed; though it be not only uſeful, but abſolutely neceſſary both to Animals, and Vegetables; without which, neither of them live, or grow, and by the means of which, the Circulation and Volatization of the blood in Animals is promoted: By the help of which, alſo the motion of every part is performed. It alſo doth not only

afford a convenient help to the Vegetation of Plants, by its compressing the surface of the water, and so forcing it to ascend into the stringy Roots and Fibers of Trees and Herbs; but also by acting the part of a Separator, [for it is, contrary to the received opinion of the *Aristotelians*, a very dry and tenious Body,] it, in its passage over the surface of the water, inbibes and takes into its Cavities, store of water, which it Transports to distant places [where Springs and Rivers are wanting;] and then being no longer able to suspend it, by reason of its plenitude, and weight, it returns it to the Earth, where it proves a fit nourishment for Plants, and a proper matter for all sort of Seeds to form themselves Bodies out of.

149 An other use of the Air, is to be a receptacle, to receive vapours ascending from the water, through the pores of the Earth, where finding many Cavities, these vapours rove about, till by the cold of the place, or the great extencion of them, the Seminal Principle contained in them, and by which they were specifically distinguished from

from water, is forced to desert the Body of the vapour; and so at last it returns to the Earth, in the form of the Catholike and universal matter, *water*.

150 It likewise serveth as a fit Body for the Stars to glide through, and move in; and also by its Elatery Spring, pressing equally upon all parts of this Terraqueous Globe, it keeps it firmly supported in its place; and doth the same Office, which I suppose *Zoreastes* means by his *Prestor*.

151 These are some of the Offices, and Uses, that God and Nature hath designed the *Expansum*, or *Firmament*, or *Etherical* Air for, but that Air we live in, and enjoy, is very far estranged from the nature of pure *Ether*, it being filled and defiled, with the Subtil steames and effluviums of all sorts of Bodies, which are there in a constant Flux, by which means particles of matter differently figur'd, [and as yet retaining some slight touch, as I may say, of their seminate natures,] meeting together, by their action and reaction upon each other, generate Metors; which having spent themselves, return to the bosome of the catholick matter, *water*.

152 But before I take leave of this ſubject, give me leave to take notice of a great miſtake in the *Ariſtotelians*; who affirm, that Air may be Tranſmuted into water; which change was never yet performed, either by Nature or Art. For, if it be to be done, by their own confeſſion it muſt be performed by the means of compreſſion, or condenſation. But compreſſion will not do the feat, as is manifeſt by winde-Guns; in which the Air is forcibly compreſſed [into, ſomtimes the Twentyeth part of the ſpace it poſſeſſed before;] yet for all that, it is ſo far from being Tranſmuted into water, that by the help of this Compreſſion, it hath its Elaſtick or Springy faculty ſo far advanced, that it will with as much impetuoſity and vigour throw forth a Bullet, as Gunpowder ſet on fire would do.

153 Nor will condenſation ſerve the turn. For the moyſture which we ſee affix it ſelf to the walls of Cellars, and Caves, or any other ſubteranious places, is not Air Tranſmuted; but the vapours of water lodged in the Cavities of the Air; which being compreſ-

ſed

ſed by the cold of thoſe places, becoms drops too bigg, and heavy for the Air to keep up; and ſo falling down, they ſettle in their priſtin ſhape of water.

154 And as Air is not Tranſmutable into water, neither is water into Air. For it is manifeſt in diſtillations, that though water be converted into very ſubtile vapours, yet by the touch of the cold Air, it returns again into water as before, and ſo diſtils into the Receiver. And I have ſhewed above, that in natures Circulations, though water be ſo diſtended as to become a moſt ſubtile vapour, or *Gas*, it doth yet conſtantly at laſt return, in its own Shape, to its own fountainwater, from whence it ſprang.

155 From what hath been ſaid, it will follow, that though we do allow Air to be a very great Body, and a conſiderable part of the Univerſe, and alſo exceeding uſeful to all Bodies, we cannot yet afford it to be a material Principle, or Element, out of which any ſublunary body is Conſtituted or Made.

156 Laſtly, let us examine whether the *Earth* have any right to be count-

ed

ed an Element or Principle, of which Bodies are conſtituted. For, although the *Ariſtotelians* [as well as the *Chymiſts*] pretend to reſolve all concretes into their firſt Principles by Fire; which they think they evince, by the example of burning wood. For, ſay they, That which ſupplies the flame, is Fire: That which ſweats forth of the ends of the wood, is water; and that which aſcends in ſmoak, is Air; but that which remaines fixed [*viz.* the Aſhes] after the Fire hath disbanded the other parts, is Earth. Yet if we examine this experiment of theirs, it will be found too Groſs, to make out what they endeavour to Illuſtrate by it.

157 For firſt, the Phlegme of the wood is not a ſimple water; but contains a ſower Salt, and doth both need, and will admit of a further diviſion to reduce it to Elementary water.

158 Nor were thoſe parts which are converted into flame, Fire; but Roziny, or [as the Chymiſts phraſe it] Oyly, or Sulphury parts: which I have before ſhewed to be far from an Elementary ſimplicity.

Neither is the ſmoak, which is ſeen

to arise in the conflagration, Air. For it will affix it self to the funnel of the Chimny in the form of Soot; after which it may be divided into Water, Oyl, Salt, and Earth, [as they call it.]

159 And the Ashes [which they are pleased to take the liberty to call Earth] every Wash-maid knows, are far enough from being so; since they are yet so compound a Body, that they contain very much of a lixiviate and fixt Salt. So that in reason it cannot be called an Element; [For Elements ought to be pure, and simple Bodies, not capable of a further reduction into different parts.]

160 And here it is necessary to remember my promise, and to take notice, that the modern Chymists, after they have washed the Salt from these Ashes, do not scruple to call it Earth, and allow it the place of one of their five Principles, of which they affirm all Bodies are compounded, and framed. But, as I declared before, so I do now again affirm, that the separating of these parts from Concrets by the force of Fire, is not a true *Analisis*, or proper way of taking Bodies to pieces; and therefore

therefore is no Genuine reduction of them; but a forcing of their parts asunder by the Fire, by which new combinations of the parts of Matter are made; and consequently the products of the Fire, are not to be looked upon as Principles, which were existing in Bodies under that form, in which the Fire presents them us.

161 Besides, were Fire an adequate and proper Agent to dissolve the Texture of Bodies, and to present us with their real Principles, it would act uniformly upon all Bodies, and exhibit to us the same Schemes of matter, with certainty from all alike; which it doth not do. For [as for example] from Gold, Silver, Talk, Diamonds, Rubies, common Stones, Sand, and many other Bodies, who ever separated? not to say the four Elements, or the five Chymical Principles, but even any two of them; and yet if we may credit that worthy man *Helmont*, all these Bodies, by the operation of his *Alkahest*, are to be reduced into simple water, equal to their own weight. So that this soluent, must [from the uniformity of its operation] be allowed to be a much

mo.e

more fit instrument to discover what Bodies are composed of, then Fire alone can be supposed to be. And if we strictly examine the business, we shall find, that Earth doth not enter any natural Body, as a constitutive Principal thereof; but indeed Earth, or Ashes, may help to compose Artificial Bodies, such as Pots, and Glasses.

162 For all sorts of Earths are but various Coagulations of water, diversified by different Seeds, and Ferments, and are as much the products of water, as I have shewed Mineral Salts, middle Minerals, Stones, &c. to be. All which, as *Helmont* assureth us, are reducible to water, by his great Solvent, [the *Alkahest*] which possibly I have somwhat more reason to affirm, than I am willing to declare.

Earth I confess, to me appeareth to be the first product of the water, and is designed by nature as a firm foundation, [or Pedestal] to support the weight of Animals, Vegetables, and Minerals, and to afford proper Wombs for the water to deposite its seeds in. For the Earth produceth nothing of its self; but all things by the assistance of water,

ter impregnated with Seeds; which it depositeth in its bosome.

163 And that the Earth was the first product of the water, is confirmed by the Testimony of *Moses*, in the first Chapter of *Genesis*, at the *9th.* verse; where describing the Creation of the Earth, he says no more but this: *God commanded the water together into one place, and the dry Land appeared.*

164 From what hath been said, it is I think, very clearly made out, that Water, and Seeds, are the true and only Principles, of which all Bodies are made, and that neither the *Tria Prima* of the old Chymists, nor the five Principles of the Chymists of our Age, no nor yet the four Elements of the *Aristotelians*, can rationally be allowed to be the Principles, or Elements of Bodies. So that as *Helmont* sayes, *ruit totum quaternarium Elementorum prætor aquam: The whole Doctrine of the four Elements falleth to the ground: Excepting water only.*

165 Having now in some measure, made out the truth, or at least probability of these Principles I assumed to defend, both by reason, and experiment; it

it remains, that according to my promiſe, I ſtrengthen theſe aſſertions by Authority. And ſhew this is no Noval opinion; but that it was held, and believed by the Antienteſt Philoſophers: Such as *Moſes*, *Sanchoniathon*, *Mochus*, *Orpheus*, *Thales*, *Pithagoras*, *Timæus*, *Locrius*, *Plato*, &c. After which I ſhall make ſome ſhort examination of the Hiſtories of Petrification, alledged in the firſt Section of this Diſcourſe, and ſo put an End to this Eſſay.

Section the Eighth.

---

166 THat *Moſes* held water to be the Firſt and univerſal Matter, will appear from what he tels us in the Firſt chapter of his Book of the Creation, called *Geneſis*, verſe the Second, where he acquaints us, that the firſt material ſubſtance out of which God made this Beautiful and Orderly frame of the World, which from its Beau-

Beauty the *Greeks* call κόσμος, was water. His words are these; *And the Spirit of God moved upon the Face of the Waters.* Where it is to be observed, that the word which our Translation renders *moved*, is in the Original *Hebrew* מרחפת, *Moracephet*; which properly signifieth not a bare motion, but such a motion as we call Hovering, or Incubation, as Birds use to do over their Eggs to hatch them. By which expression we have not only an account of the first matter out of which the World was afterwards made; but also of the Efficient, by which this matter was wrought into so great a variety of Bodies. For in all probability, the sence of the Expression is, that at that time, [*viz.* in the beginning] God infused into the bosome of the waters, the seeds of all those things, which were afterwards to be made out of the waters, setting them their constant Laws, and Rules of acting [and thus was Nature Created, that is, the Order, and Rule of those things were established, which God designed to make:] and by the power of the words, *increase and multiply*, they had a facul-

ty

ty given them, to continue themſelves in the ſame Order, till the world ſhall be deſtroyed by Fire, [the great deſtroyer of Seeds;] at which time all Seminal beings ſhall deſert their groſs Bodies, and return to their firſt Fountain, and great exemplar God, on whom they have at this time a conſtant dependance. For according to the Apoſtle, *In him and to him, and through him, are all things; and in him we live, move, and have our being.*

167 *Sanchoniathan*, the great *Phenician* Philoſopher, [whom ſome Chronologers make contemporary with *Gideon*] ſome part of whoſe Works are yet to be met with in *Philo-Biblius*, and *Euſebius*; and a good account of whoſe Works we may alſo find in the writings of that Learned, and Ingenious man, Mr. *Gale*. This *Sanchoniathan* I ſay, exactly correſponds with *Moſes*. For he ſays, In the beginning there was χάος ἐρεβῶδες, which in the *Phenitian* Tongue, is כוות ערב *Chauth Ereb*; that is, *Night or Evening Darkneſs*. Then he further ſayeth to this purpoſe; *From the commixtion of the Spirit with the Chaos, was produced* *Court of the Gentils. part 2d. p. 5*

Mot, which ſome call [ἰλὺν] that is *matter, or watery moyſture: Out of this was produced the whole Seed of the Creation, and the Generation of the whole.*

168 Alſo *Mochus*, an other *Phenizian*-Philoſopher, who continued the Philoſophick Hiſtory, begun by *Sanchoniathon*, [and who is ſaid to have written long before the *Trojan* War,] was alſo of the ſame opinion, as *Bochard* affirms.

169 And that *Thales* of *Miletus*, [who is held the firſt Philoſopher that writ in *Greek*] taught that the world was made out of water, no body can be ignorant. And that, which *Sanchoniathan* calls *Mot*, fluid Matter, he calls ὕδωρ, water. And *Tully* affirms, that *Thales held water to be the begining of things: And that God out of water framed all things.*

Tully *de Natur. Deorum. lib.* 1. *cap.* 2*d.*

170 *Orpheus* alſo is of the ſame judgment, and tells us, ἐκ τοῦ ὕδατος ἰλὺς κατέστη; *of water, Slime was made.* And *Apollonius* ſays, ἐξ ἰλύος ἐβλάστη χθὼν αὐτή. *Earth, of Slime was made.* And the Scholiaſts give a good explication of theſe words; for they affirm, that the *Chaos,*

Chaos, of which all things were made, was water, which coagulated it ſelf, and became Slime; and that Slime condenſed, became ſolid Earth.

171 Thus you ſee, that *Thales*'s ὕδωρ, or water; and the χάος, μῶτ, and ἴλυς, *i. e.* watery moyſture, of *Sanchoniathon*, and *Mochus*, was believed and held by them to be the firſt Principle of all things: From which the ὕλη of *Pythagoras*, and *Plato*, differs not; as I will ſhew by and by.

172 *Pherecides* [an antient *Greek*-Philoſopher] who was *Pythagoras* his Maſter, and who we are told, was one of the firſt *Greeks* that held the Immortality of the Soul; though he ſeem to differ from *Thales*, and *Orpheus* in ſome things, yet agreed with them in the main, or the thing taken for granted by them all, *viz.* That *water* was the firſt Matter of all things.

173 Alſo *Pythagoras*, the Founder of the *Italick* Sect of Philoſophers, correſponds exactly in Opinion with *Moſes*, concerning the Origin of the World, and its firſt Matter. For he poſitively held, that the World was made by God; and by him adorned with an

excellent Order, Harmony, and Beauty in all its parts; and therefore he was the first that called it κόσμος, from κοσμεῖν, to Adorn or Beautify: Secondly, his ὕλη, or first Matter, was the same with *Sanchoniathons* ἰλὺς, or Μῶτ; and *Thales* and *Orpheus* their ὕδωρ, *viz.* water: Agreeable all of them to *Moses*, *Genesis* the first.

Thirdly, *Pythagoras*, and all the Antient Philosophers before him, held, that the Divine Providence, which they stile νοῦς, did inspire and influence the whole Creation, governing, and directing all things to their proper and peculiar Offices, Functions, and Ends. And this Providence was by them somtimes stiled ψυχὴ τοῦ κόσμου, the Soul of the World; by which, sayth *Seranus*, they understood nothing else but the Fire, Spirit, or Efficacy, which is universally diffused in the Symmetry of the Universe for the Forming, Nourishing, and Fomenting all things according to their respective natures: Which Vivifick Principle *Plato* calls πῦρ δημιουργὸν, effective Fire; but this they never understood, or meant to be a material part of any Body; but is the same which

Mo-

Moſes calls the Spirit of God.

174 And now in the laſt place, I am come to give you the mind of *Plato*, and his conformity with *Moſes*; His judgment hath always been ſo eſteemed, that men, to expreſs the Reverence they had of him, did uſually call him the Divine *Plato*: And in delivering his opinion, I ſhall alſo at the ſame time give you that of *Timæus Locrius*, that great Philoſopher, and Diſciple of *Pythagoras*; from whom *Plato* borrowed much.

175 Firſt then, *Plato* tells us that the World was made: For he puts the queſtion whether the World had a beginning, or was made? To which he anſwers, γέγονεν, it was made. Then as to the firſt matter, of which the World, and all the Bodies in it were made, he ſays thus, [in his *Timæus*] it is γένος, εἶδος ἐξ οὗ τὸ πᾶν συνεςέθη, *the Genus or Species out of which every thing is compoſed*; and He calls it ὕλη, or firſt Matter, and is indeed the ſame with *Sanchoniathans* ἰλὺς, *Mot*, &c. and *Thales*, and *Orpheus*'s ὕδωρ; and all of them the ſame with *Moſes* his *Chaos* and *Water*, as will appear by comparing

their descriptions together. Thus first, *Moses* calls his first matter בהו *Bohu*, without form; which *Rabby Kinchi* calls ὕλη [as *Fabius* tells us] which is the same word that *Plato* uses to express his first matter by; and differs little, in sound, but less in the sence from the ἰλὺς of *Sanchoniathon*, which *Philo Biblius* stiles *Mot*, from the *Hebrew*, and *Phenitian* מוד *Mod*, which signifieth Matter: Yea, *Plato* expresly calls his first Matter ἀμορφόν τι, *somewhat without form*; just like *Moses* his *Bohu*.

176 And in his *Timæus* he tells us, that God out of this first matter [*water*] commonly called *Chaos* [because disordered, and irregular] διεκόσμησε, διέταξε, καὶ διεσχηματίσατο, *Beautified, Ordered, and Figured, or Form'd the Universe*; and as *Moses* says, the Spirit of God moved upon the Face of the waters: So *Plato* affirmeth, that God made the World, ἐκ ἡσυχίαν ἄγον ἀλλὰ κινέμενον πλημμελῶς καὶ ἀτάκτως, that is, *by an importunate motion, fluctuating, and not quiescing upon the matter*. And as for *Plato*'s ψυχὴ τῦ κόσμε, or *Soul of the World*, we are assured by *Ludovicus Vives*, he meant by it the same Spirit of God which *Moses*

Lud. Vives in com super

Moses says moved upon the waters in the Beginning; and which the Psalmist calls the breath of his mouth: (*Psalm* 33. *verse* 6.) For, according to *Platoes* Philosophy, [as well as that of *Moses*] God is the Executive cause, and productive Efficient of all things, and therefore he usually stiles God, ἀρχηγὸς, πρωτεργὸς, τελεσιουργὸς, ὀυσιοποιὸς πάντων ὄντων, *the Supream Fabricator, Perfector, and Essentialisor of all things.* And as to the manner, how all things were made, he says, Σπαστηρίοις λόγοις τὸ πᾶν ὀυσιῶται; *Every thing was essentialised by certain Prolifick, or efformative words,* which the *Stoicks* call λόγον σπερματικὸν, a Spermatick, or Seedy word: Which agrees exactly both with *Moses* his *Fiat*, and with that of St. *Paul*; *The Worlds were framed by the word of God*; that is, Gods *Fiat* was the Creator of all the Seminal and Prolifick Principles of all things; and those created Seeds were τὸ ποιητικὸν, *the Efficients*; and ἡ ὕλη, or ὕδωρ, *water*, was the Matter of which they were all made.

Epistle to the Heb. cap. 11. verse 3.

177 These Seminal or Efficient Principles of things do contain within them-

ſelves certain Pictures or Images of thoſe things which they are to make out of the matter, [*viz. water.*] To which purpoſe let us here what *Plato* ſays of his *Ideas*, which is to this effect; *There are two ſorts of Worlds; one, that hath the form of a Paradigm, or Exemplar, which is an intelligible Subject, and always the ſame in being: but the ſecond, is the Image of the Exemplar, which had a beginning, and is viſible.* By his Intelligible World, *Plato* means the Divine Decrees; which are inherent in the Mind and wiſdom of God: and theſe Original *Idea*'s, he ſays, do produce a Secondary ſort of *Idea*'s [that is, the *Seeds* of things;] and theſe he makes to be the more immediate Delineation, or Image of the whole work; ſomtimes calling them παράδειγμα, *an Exemplar*; ſomtimes εἰκόνα, *an Image*: His words run thus; τοιέτῳ τίνι προσχρώμενος παραδείγματι τὴν ἰδέαν καὶ δύναμιν ἀπεργάζεται: *making uſe of this Exemplar, he frames the Idea, and Powers*; that is, *the Seeds of things.* So that he makes the firſt, and Original *Idea*, [which is reſident in the Divine Wiſdom or Mind of God, and which

Plato Timæus. fol. 49.

which Divines call the Decrees of God] to be much more Noble than the latter, or secondary *Idea*, or Seed, and to be the cause of it. And this last *Idea* and Seed, contains the Picture of the thing to be made; and depends upon the Primary, or Original *Idea*, and Exemplar, which is seated in God himself.

179 Which Doctrine rightly considered, we have a satisfactory account of the cause, why the last *Idea's*, viz. the Seeds of things do proceed so regularly, constantly, and unerringly in the producing their likes. For, if we consider, that the Seeds of things do depend upon their Paradigmes, and that they are inherent in the Mind of God himself, who is a God of Order; this will appear not so abstruse, as it hath hitherto done.

180 And though we, out of Pride, and self-love to our own Nature, are unwilling to afford any creature, that is not of our Species, the Priviledg of doing any thing by a Principle of reason; that is, with a design, tending towards the accomplishment of such an End; yet it is certain, that all crea-

tures,

tures, even thoſe that we count inanimate, do enjoy, upon the account of their Seminal Principles, not only Life, but even reaſon in ſome meaſure: Which, wanting the uſe of Languages, they do nevertheleſs plainly declare [to heedful and inquiſitive men] not only by their regular, [and conſequently deſigned] working the parts of matter, till they have produced ſuch a diſtinct ſort of Body; but alſo by thoſe affections which wee call Sympathy and Antipathy; and, for want of this knowledg, have hitherto referred to occult or hidden cauſes, the uſual Sanctuary of Ignorance; by which Sympathy, and Antipathy of theirs, it is very manifeſt, they have hatred and love; and have a knowledg of thoſe things, which are either pleaſing or agreeable too, or elſe unpleaſant or hurtful to their natures. And this is not only to be obſerved in Beaſts, and viſible moving Creatures, but alſo in all other ſorts of Creatures, which we very injuriouſly call Dead, or Inanimate.

But

181 But to return from whence I digreſſed, I ſhall in ſhort ſay thus much of the *διότι, or manner how* the Ideas and Seeds do work upon Matter, and form themſelves Bodies; which they perform on this manner: Firſt, by their Fermentative faculty, [or Springy power] they put the Body of the water into a peculiar ſort of motion, by which they congregate thoſe particles, which are moſt agreeable to their deſign, and conſequently fitteſt to adhere, and ſtick to each other. Secondly, they break the reſt into convenient ſhapes, and Sizes: And Thirdly, by this motion they alſo put theſe particles into commodious Poſtures, and Scituations amongſt themſelves, and by theſe means frame themſelves Bodies, exactly correſpondent to their own præconceived Figures.

182 By this declaration of my thoughts, I hope it will plainly appear, that I am no Enemy to that rational way of explicating the *Phænomena* of Nature, uſed by the *Atomical, Carteſian*, or *Corpuſcularian* Philoſophers; for certainly, they do

give

give us not only a very ingenious, but alſo a very true account of the διότι, or manner how, matter is, or may be modified; to which, if they would pleaſe to add, as ſome do, the powerful efficacy of Seeds upon Matter, by which indeed all the ſeveral σχέσεις, or various ſhapes of Matter is produced, we might then hope to receive ſome ſatisfactory account of that hitherto perplexed Subject, the Generation of Natural Bodies: Which Principle if it were received, and taken into the Philoſophy of our Age, I am apt to believe it would ſilence many Litigations, now daily commenced by men of Parts againſt each other; and oblige them to love truth more then the deſire of being accounted witty Diſputants; Truth being ſo deſirous a thing, that *Porphyry* in the Life of *Pythagoras* [though a Heathen] tells us; τὸ ἀληθεύειν μόνον δύναται, τοὺς ἀνθρώπους ποιεῖν θεῷ παραπλησίους, that is, *Truth only can make men near to God.*

183. Now therefore, though rude and unguided motion, will naturally have ſome kinds of reſult upon matter, as

as we ſee the ſpringy motion of the Air, or ſome more ſubtile Body doth form of the Water, of Rain, and Dew, round Drops, by equally Compreſſing it; yet because this general kind of motion doth ſomething, we are not from thence to conclude it doth all things. For, this were a Sophiſme, fitter to impoſe upon Fooles or Children, then upon Men of mature reaſon. Nor can ſuch kind of motion be ever able to forme ſuch bodies, as imply a wiſe Couńcel, and curious contrivance; as, for Example [to ſay nothing of Living Creatures] the ſtrong and uſeful bodyes of Metals, Minerals, and Stones, and the beautiful Branches, Flowers, and Fruits of Plants, are. Wherefore we muſt in all reaſon acknowledge and confeſs, that there is an internal Mind, virtue, and Idea, contained in the Seeds of things, which workes rationally, [that is, to a Deſigned end;] by which Principle, the matter is put into a peculiar motion, and uſefully guided, till it be changed, and formed into a body, ſuch as the Idea was deſigned by God to make, who ſtill governes theſe Seedy Principals: And therefore in

Scrip-

Scripture, we are told, *He Giveth to Every Seed, its own Body.*

184. Thus then, I hope, I have proved, that I am of the same Judgement with the Antientest, and best Philosophers; viz. that there is but Two Principles of all things, Efficients, and Matter; Seeds, and Water.

185. And now having cleared the Doctrine proposed; I intend in the last place, to inquire, How those Transmutations of different Bodies into Stone, the Historyes of which you will find set down in the first Section of this discourse, were performed: upon which, I will only Touch, and so Conclude.

186. It is the Opinion of some Men, that the change of Leaves, Mosse, Wood, Leather, and other Substances, into Stone, [wrought by those Petrifying Waters, and Caves, I have mentioned in the first Section of this Essay] are no real Transmutations of those Bodies into Stone, by the Operation of a Petrifying Seed; but that they are nothing else, but the opposition of certain small Stony Particles, hid in the Water, to those Bodies immersed in them; and that by this means they become Crusted over with

with a stony Coat or Bark, and so they become increased both in Bulk, and Weight, by continual addition. But if this were so, then indeed the Leaves, Wood, &c. cast into these Waters, would not be really transchanged into perfect Stony Nature; but only seemingly so.

187. Neverthelefs, if we look warily into the thing, we shall have Cause to believe, that there is, not only an Aggregation of these small Stony particles, and an incrustation upon the outside of those things put into the Water; but even that the smallest Atomes of the Wood, Leather, &c. are really Petrifyed; in so much, that we can discern them to be no other then Stones, not only by our Eyes alone, but by them assisted with the best Microscopes. Nor if they be examined by the Fire, will they make any other Confession: For they will not burn like Wood, but calcine like Stones; and though great peices of Wood, and Trees, will not be so soon converted into Stone, as Twiggs, Leaves, or Moss, are; yet by continuance of Time, great bulkes of Wood will be Stonifyed totally, both within, and

and without; so that by these kind of Waters, bodies are not only Crusted over with stone, but the Wood, Leaves, &c. are really and truly changed into Stone. I do not deny, but that there may be an affixing of some stony Corpuscles Latent in these Waters, which may increase both the bulk and weight of those things Changed by them; but that this is all, that I deny.

188. For, if so, then those Bodies thus changed, would not be altered into a true Stony Nature, *per minima*, and in their smallest parts, internally, as Experience shews they are; and though the Explicating, how this Change is Wrought, is somewhat difficult, yet in all probability it is thus.

189. The Saxeous, or Rocky Seed, contained in these Waters, [which is so fine, and subtile a Vapour, that it is Invisible; as I have before shewed all true Seedes are,] doth penetrate those Bodies which come within the Sphere of its Activity; and by reason of its Subtilty, passeth through the pores of the Wood, or other Body, to be changed: by which permeating those Bodies, it doth these four things: First,

it Extruds the Globuli Ætherai [as the Cartesians Phrase it] or the Airy Particles Lodged in their pores: Secondly, it puts the Particles of those Bodies into a new and different motion, from that they were in before; by which meanes they become broken into Figures, and Sizes, and obtain new and convenient Situations. Thirdly it intangleth and Lodgeth it self intimately amongst the smallest parts of those Bodies; by which meanes their parts being drawn closer together, they obtain a greater Weight and Solidity: And lastly, it Acts as a Ferment, and by reason of its Contiguity, and Touch with every small part of the matter it doth, as Leaven useth to do, [though mixed with a much greater quantity of Dough, then it self] Convert the whole into its own Nature. So also this Stonifying Seed, by its operating Ferment, doth transchange every particle of the matter it is joyned unto, into perfect Stone; according to its Idea or Image, Connatural with it self.

190. As to those Conversions of Animals into Stone, related in History, the 13, 15, 16, and 17, of the first

Section of this Essay; they also are wrought by the same powerful Operations of a petressent Seed or vapour; and by the same Circumstances, and Contrivances: which sheweth, that the strength and Power of a Petrifying Seed is above, and beyond all other: For, other sorts of Seeds do require, that the subject matter be reduced into a sequatious juice, or obedient Liquor, and Consequently doth require, that the Figure, and Shape of the precedent Concrete be destroyed, or else they cannot Act. But the Petrifying Seed, the Human, or other Living Creatures Figure being still intire, without any intervening putrefaction, or dissolution of the matter, doth transchange [*Totum per Totum*] the whole, throughout the whole; that is, as well the Bones, as the Blood, and Skin: So that here is not an incrustation of the Stony matter upon the External parts, [only] but a real change, intrinsically, and throughout, of the Bony, Fleshy, and Sinnewy parts of the Animal into a stony Substance.

191. By the same operations Water it self is converted into Stone, [*viz.* by

by, the power of Petrifying Seeds] as we may ſee by the 7, 8, 9, 10, 11, 12, and Fourteenth Hiſtory of the firſt Section: As alſo doth appear by the Relation of thoſe that have ſeen thoſe Famous Grots in *France*, called, *Les Caves Goutieres,* where the Drops falling from the top of the Cave, doth [even in its falling] coagulate before your eyes into little Stones. Now this Tranſmutation of Water into Stone, by a Petreſcent Seed, is not only much more uſual, than the change of other Subſtances is, but alſo much Eaſier: For Water is a Primary, ſimple, liquid, tremulous Body, conſiſting of very minute parts, already in Motion, and therefore readily Obeying the Command of all ſorts of Seedes.

192. Nature is uniforme in her manner of produceing Bodies, and therefore, as I have demonſtrated in the body of this Diſcourſe, as ſhe uſually, nay conſtantly produceth, both Animals, Vegetables, and Mettals, from liquid Principles, viz. Water. So doth ſhe moſt commonly Stones; which before their becomming ſuch hard Bodies, were at ſometime in *Principiis*

Solutis, that is, their matter was in a loose, open, and fluid Forme: And, as I have shewed, the Spiritual Seedes of Vigetables, do assimilate, and change Water, into Mint, Rosemary, &c. According to the diverse Ideas, and characters of their peculiar Kindes; so also the Stony Seedes, do form themselves Bodies out of Water; and these of very different Figures, Compaction, and Colours; and this is done sometimes suddenly, sometimes slowly, and by length of time: Now, the difference of compaction, and hardness, that we find in Stones, as also their sudden or slow Coagulation, depends chiefely upon the plenty, or paucity of the stony Seed, or Spirit, in respect to the quantity of the matter to be wrought upon, and changed by it. But the difference of the Figure, is chiefely to be referred to the peculiar Nature of the Seed, and its Idea; [as we see in Christals, and other Stones, which have a determinate Figure:] and sometimes it is to be referred to the vessel, or place, containing the Water, or other Liquor, before its conversion into Stone. And for the Colour, that is also chiefely caused

ſed by the operation of the peculiar ſort of ſtony Seed; which in its working upon the Water, hath given it a determinate Texture, and ſuperficies; by which it reflects and modifies the Light, after a peculiar manner. But ſometimes it is to be referred alſo to the Waters being impregnated with the Tinctures of ſome Mineral or Mettallin Bodies, before its coagulation. As Granets containe the Tincture of Iron in them; and therefore are drawn by the Loadſtone.

But to put it out of all doubt, that Stones were at firſt Water; [or at leaſt, ſome Liquid Matter] I will Cite a paſſage or two out of the Works of my often mentioned, Honourable Friend, Mr. *Boyle*. His words are theſe: *And here I will Confeſs further, that I have oftentimes doubted, whether or no not only Conſiſtent Bodyes, but ſome of the moſt Solid Ones in the World, may not have been Fluid in the form, either of Steemes, or Liquors, before their Coalition and their Concretion either into Stones, or other Mineral Bodies.* And then ſpeaking of the Opinion of ſome Men, who will have it, that Stones, and Mettals, [were

Boyl in his Eſſay of Ferm. p. 281.

 indeed

indeed Created at the beginning of the World by God, but that since they] are neither Made, nor do Grow, and increase: He further saies [viz. that they were once in a fluid forme] thus: *Of this, besides what we elsewhere deliver Concerning it, we shall anon have Occasion to mention some Proofs; and therefore we shall now only mention two or three instances: the first whereof shall be, that we saw, among the rarities of a Person, exceedingly Curious of them, a Stone flat on the outside, on one of whose internal surfaces was most Lively Ingraven, the Figure of a small Fish, with all the Finns, Scales, &c. which was affirmed to have been inclosed in the Body of of that Stone, and to have been accidentally discovered, when the Stone chancing to receive a rude Knock upon its Edge, split a sunder. I Remember also that a while since a House-keeper of mine in the Countrey informed me, that whilest a little before, he Caused in my absence one of my Walls to be repaired; the Mason, I was wont to imploy, Casually breaking a Stone, to make use of it about the Building, found in it [to his Wonder] a peece of Wood, that seemed part of of the branch of some Tree, and Consequently was afterwards inclosed with that solid Case wherein he found it. This Example seemes to me a mere cogent Proof of the increase of Stones, then some others, that Eminent Naturalists much rely on, for reasons discoursed of in an other place.*

193. And again, He tells us in the same place, that He hath seen several large Stones, such as they make Statues of, that when they were sawed, and broken, had Caveties in them, which contained Mettals, and other substances: And I my self have observed pebbles inclosed in great free Stones. And it is commonly known, that Spiders and Toads have been found upon the breaking of great Stones, inclosed in their innermost substance.

194. And now I have shewed you, how agreeable I am with this Learned Person in this Doctrine concerning the matter, and growth of Stones; I will also shew you his Opinion, as to their Efficient: for he says; *I know that not only profest Chymists, but other persons who are deservedly ranked amongst the Modern Philosophers, do with much Confidence* Essay of Ferm. p. 275.

entirely ascribe the induration, and especially the Lapidescence of Bodies, to a certaine secret internal principal, by some of them called a Forme, and by others a Petrifying Seed, lurking for the most part in some Liquid Vehicle: And for my part, having had the opportunity to be in a place, where I could in a dry Mould, and a very elevated peice of Ground, cause to be digged out several Christalline Bodies, whose smooth sides, and Angles, were as Exquisitly figured, as if they had bin wrought by a skillful Artist at cutting of precious Stones; and having also had the opportunity to consider divers exactly or regularly shaped Stones, and other Minerals, some digged out of the Earth by my Friends, and some yet growing upon Stones, newly Torn from the Rocks, I am very forward to grant that [as I elsewhere intimate] it is a Plastick Principal implanted by the most wise Creator, in certain parcels of matter, that doth produce in such Concretions, as well the hard Consistance, as the determinate Figure. Thus far He; Then which, what more consonant to the Doctrine I have asserted in this Discourse?

195. Conclude we then [and I hope at

at last upon probable Grounds] since we have not only the before cited Authorities, both of the best Antient, and Modern Philosophers; and also are taught by the experiments, and Manual Operations laid down in this Discourse, which shew us the reduction of all bodies ultimately into Water; and their Nourishment from thence; as also from the inaptitude of at least two of the four *Aristotalian* Elements [*viz.* Fire, and Aire] to concur to the Constituting of Bodies; and likewise from the Compound Nature, of two of the Old Chymical Principles, viz. Sulphur and Salt: and from the same compound Nature of four of our moderne Chymists Principles, viz. Oyle, Salt, Spirit, Earth, which all of them are further reducible into Water, and therefore not to be allowed for Principles; as I have before demonstrated: Let us then, I say, conclude in, and acknowledge the truth of the Moysaick, Platonick, and Helmontian Doctrine.

195. That is, that all Bodies consist but of two Parts, or Priniciples, Matter, and Seed; that their Universal Matter is Water: That the Seedes of things do from

from this Matter, [by the help of Fermentation] alter, break, and new compose the Particles of which it Consists, till they have formed a Body, Exactly Corresponding to the Images, or Idea's contained in themselves: Also that the true Seedes, of all things, are of a very subtle Nature, and Invisible, and are secundary Idea's and Images; and that they are Connexed to, and depend upon their Primary Idea's, and Exemplars; which are Inherent and resident in God himself: And that for that reason they Act with Designe, and to a purposed End, which they constantly, and regularly Accomplish; and this is somewhat Analogous to reason in them. Lastly, that Nature, or the Law of Kind, is uniforme in its productions thus far, that it makes all Bodies out of Water, by the power of invisible Seedes; so that the Matter of all Bodies is Identically the same. And that they are all of them reducible into the same Matter at Last: But that their Seeds are various, and therefore produce different Effects upon the same Matter: yet do they all agree in this, viz. That they are all invisible Beings, and all of them have a dependance upon on their Exemplars, which are the Decrees of God, and are conſtantly inherent in him.

FINIS.

An Advertiſement.

THere is lately Printed, a Book, *in which is ſhewed the neceſſity that lies upon all Honeſt, Diſcreet, and Conſcientious Phyſitians, to reſume that Antient, and Laudable Cuſtom of making, and Diſpencing their own* Medicines; *with the Advantages thereby accrewing to the* Patient: *Both as to ſaving of Charges; and the ſpeedy cure of their* Diſtempers. *In which the New way of Preſcribing* Bills, [*or making Medicines with the* Pen] *is ſhewed to be deſtructive to the Intereſt, both of the* Patient, *and* Phyſitian: *It expoſing them to the Fraudulent dealing of Practiſing* Apothecaries, *in which you will find the Marrow of what hath been writt upon this Subject, by Dr.* Cox, *Dr.* Merrit, *Dr.* Goderd, *and others; together with certain new, and cogent Arguments not formerly*

merly made use of. The Subject I conceive, of such general concern, that I thought it it very fit to give notice of it here.

The Title of it is Praxis Medicorum Antiqua, & Nova, *Or the Ancient, and Modern Practice of* Physick *examined, Stated, and Compared,* &c. *It was written by the Industrious, and Ingenious Dr.* Everrard Manewring. *And is to be sold by* William Cademan *Bookseller, at the Sign of the* Popes Head, *at the little Door of the* New Exchange, *next* Durham Yard.

Clarks *Examples in two Volumes* in Fol.
Bacons *Natural History* in Fol.
Reynolds *of Murther* in Fol.
Cozens's *Devotions* in 12.

Playes.

Cambyses *King of* Persia *in* 4.
Island Princess in 4.
Town Shifts in 4.
Juliana *in* 4.
Cataline *in* 4.
Rivals in 4.
Flora's *Vagaries* in 4.

Mar-

Marcelia *in* 4.
Imperial in 4.
Fortune by Land and Sea in 4.
Unfortunate Mother in 4.
Hamlet *in* 4.

Cum multis Aliis.

To be Sold by William Cademan, *at the Signe of the* Popes Head *in the* New Exchange.

History of Geology

An Arno Press Collection

Association of American Geologists and Naturalists. **Reports of the First, Second, and Third Meetings of the Association of American Geologists and Naturalists, at Philadelphia, in 1840 and 1841, and at Boston in 1842.** 1843

Bakewell, Robert. **An Introduction to Geology.** 1833

Buckland, William. **Reliquiae Diluvianae:** Or, Observations on the Organic Remains Contained in Caves, Fissures, and Diluvial Gravel. 1823

Clarke, John M[ason]. **James Hall of Albany:** Geologist and Palaeontologist, 1811-1898. 1923

Cleaveland, Parker. **An Elementary Treatise on Mineralogy and Geology.** 1816

Clinton, DeWitt. **An Introductory Discourse:** Delivered Before the Literary and Philosophical Society of New-York on the Fourth of May, 1814. 1815

Conybeare, W. D. and William Phillips. **Outlines of the Geology of England and Wales.** 1822

Cuvier, [Georges]. **Essay on the Theory of the Earth.** Translated by Robert Kerr. 1817

Davison, Charles. **The Founders of Seismology.** 1927

Gilbert, G[rove] K[arl]. **Report on the Geology of the Henry Mountains.** 1877

Greenough, G[eorge] B[ellas]. **A Critical Examination of the First Principles of Geology.** 1819

Hooke, Robert. **Lectures and Discourses of Earthquakes and Subterraneous Eruptions.** 1705

Kirwan, Richard. **Geological Essays.** 1799

Lambrecht, K. and W. and A. Quenstedt. **Palaeontologi:** Catalogus Bio-Bibliographicus. 1938

Lyell, Charles. **Charles Lyell on North American Geology.** Edited by Hubert C. Skinner. 1977

Lyell, Charles. **Travels in North America in the Years 1841-2.** Two vols. in one. 1845

Marcou, Jules. **Jules Marcou on the Taconic System in North America.** Edited by Hubert C. Skinner. 1977

Mariotte, [Edmé]. **The Motion of Water and Other Fluids.** Translated by J. T. Desaguliers. 1718

Merrill, George P., editor. **Contributions to a History of American State Geological and Natural History Surveys.** 1920

Miller, Hugh. **The Old Red Sandstone.** 1857

Moore, N[athaniel] F. **Ancient Mineralogy.** 1834

[Murray, John]. **A Comparative View of the Huttonian and Neptunian Systems of Geology.** 1802

Parkinson, James. **Organic Remains of a Former World.** Three vols. 1833

Phillips, John. **Memoirs of William Smith, LL.D.** 1844

Phillips, William. **An Outline of Mineralogy and Geology.** 1816

Ray, John. **Three Physico-Theological Discourses.** 1713

Scrope, G[eorge] Poulett. **The Geology and Extinct Volcanos of Central France.** 1858

Sherley, Thomas. **A Philosophical Essay.** 1672

Thomassy, [Marie Joseph] R[aymond]. **Géologie pratique de la Louisiane.** 1860

Warren, Erasmus. **Geologia:** Or a Discourse Concerning the Earth Before the Deluge. 1690

Webster, John. **Metallographia:** Or, an History of Metals. 1671

Whiston, William. **A New Theory of the Earth.** 1696

White, George W. **Essays on History of Geology.** 1977

Whitehurst, John. **An Inquiry into the Original State and Formation of the Earth.** 1786

Woodward, Horace B. **History of Geology.** 1911

Woodward, Horace B. **The History of the Geological Society of London.** 1907

Woodward, John. **An Essay Toward a Natural History of the Earth.** 1695